HOMELANDS

CREATING THE NORTH AMERICAN LANDSCAPE

Gregory Conniff, Edward K. Muller, David Schuyler · *Consulting Editors*

George F. Thompson · *Series Founder and Director*

Published in cooperation with the Center for American Places
Santa Fe, New Mexico, and Harrisonburg, Virginia

HOMELANDS

A Geography of Culture and Place across America

Edited by Richard L. Nostrand and Lawrence E. Estaville

The Johns Hopkins University Press · Baltimore and London

9 8 7 6 5 4 3 2 1

The Johns Hopkins University Press
2715 North Charles Street
Baltimore, Maryland 21218-4363
www.press.jhu.edu

Library of Congress Cataloging-in-Publication Data
Homelands : a geography of culture and place across America /
edited by Richard L. Nostrand and Lawrence E. Estaville.
     p. cm. — (Creating the North American landscape)
"Published in cooperation with the Center for American Places,
Santa Fe, New Mexico, and Harrisonburg, Virginia."
Includes bibliographical references (p.   ) and index.
   ISBN 0-8018-6700-2 (hardcover : alk. paper)
1. United States—Geography.   2. United States—Civilization.
3. Human geography—United States.   4. United States—Description and travel.   5. Regionalism—United States—History.
6. Place (Philosophy).   7. Pluralism (Social sciences)—United
States—History.   8. Ethnicity—United States—History.
9. Immigrants—United States—History.   10. Minorities—
United States—History.   I. Nostrand, Richard L. (Richard Lee),
1939–   II. Estaville, Lawrence E. (Lawrence Ernest)
III. Center for American Places.   IV. Title.   V. Series.
   E161.3 .H66 2001
   304.2'0973—dc21

                                              00-012362

A catalog record for this book is available from the British Library.

CONTENTS

LIST OF MAPS, FIGURES, AND TABLES

Maps

Figures

Tables

ACKNOWLEDGMENTS

This present volume contains the thoughts of many geographers who agree with us that the concept of the homeland needs our attention. We are much indebted to our authors, all geographers, who made possible this first extended yet far-from-definitive treatment of the concept of the homeland in the United States. In preparing this volume for publication we thank Lisa M. DeChano and Julie Henry of Southwest Texas State University for their technical assistance in constructing maps. And we thank Susan E. Nostrand of the University of Oklahoma for her hours spent formatting the entire manuscript twice.

Daniel D. Arreola thanks the Hispanic Research Center at Arizona State University for travel assistance to San Ygnacio in 1993 and San Antonio in 1994. He also thanks the National Council for Geographic Education for permission to reproduce portions of *Mexican Texas: A Distinctive Borderland*, published in Pathways in Geography, Resource Publication no. 12, 1995: 3–9, and the American Geographical Society for permission to reproduce portions of "Urban Ethnic Landscape Identity," published in *Geographical Review* 85, no. 4 (1995): 518–34.

Lowell C. "Ben" Bennion warmly thanks the following: Mary Beth Cunha, manager of Humboldt State University's (HSU) Kosmos lab, for drafting his maps; Professor Philip Barlow, co-editor of *The New Historical Atlas of Religion in America*, for permission to modify his version of map 12.1; Professor Emeritus D. W. Meinig for permission to modify his version of map 12.3; and former HSU student Rena Christensen for conducting the survey of Utahns' attitudes toward Mormon Country that he draws upon.

Lawrence E. Estaville's chapter draws particularly from his article, "Changeless Cajuns: Nineteenth-Century Reality or Myth?" *Louisiana History* 28, no. 2 (1987): 117–40.

Susan W. Hardwick's chapter draws on her prior publications on Russian settlement in California, especially *Russian Refuge* (Chicago: University of Chicago Press, 1993). She thanks Lawrence E. Estaville for his support of her research on this immigrant group and other efforts to document ethnic settlement in Texas and California.

Stephen C. Jett thanks David M. Brugge, Charlotte Johnson Frisbie, and Lisa Roberts Jett—as well as the editors of the present volume—for valuable comments on a draft of his manuscript; Teresa L. Dillinger for bringing a reference to his attention; and Douglas Van Lare for creating his map.

Ary J. Lamme III thanks Mark McLean of the University of Florida Cartography Laboratory for drafting his map of the Amish homelands.

Richard L. Nostrand's chapter draws heavily on three earlier publications: Chapter 10 in *The Hispano Homeland* (Norman: University of Oklahoma Press, 1992), 214–26; "The New Mexico-Centered Hispano Homeland," *Journal of Cultural Geography* 13, no. 2 (1993): 47–56; and five maps (used in map 10.1) in "Greater New Mexico's Hispano Island," *Focus* 42, no. 4 (1992): 13–19.

Steven M. Schnell thanks James R. Shortridge, his advisor at the University of Kansas, for editorial suggestions and encouragement. He also acknowledges the many Kiowas who gave generously of their time to answer patiently his naïve questions. His chapter draws heavily from "The Kiowa Homeland in Oklahoma," *Geographical Review* 90, no. 2 (2000): 155–76.

INTRODUCTION

Free Land, Dry Land, Homeland

. . . .

Richard L. Nostrand
and Lawrence E. Estaville

The Context

Late in the evening of 12 July 1893, Frederick Jackson Turner read his seminal essay "The Significance of the Frontier in American History" (Turner 1920, 1–38). The historians assembled at a special meeting of the American Historical Association in Chicago to hear Turner's paper, the last of five read that evening, were so exhausted from the long day that the session adjourned without discussion. Only several years later did historians recognize the importance of what Turner had said. Recounting what he had read in a Census Office publication regarding the 1890 census, Turner announced that 1890 marked the closing of the American frontier. This closing was significant, Turner asserted, because it meant an end to a frontier process that had shaped the American character. Turner theorized that free land, by which he meant available land, and its continuous recession to the west engendered in Americans qualities of democracy, freedom, individualism, and nationalism. According to interpretations prevalent in 1893, the American character was little more than an overseas extension of Europe—merely European cultural baggage transplanted to America (Billington 1973, 117–18, 127, 129, 184, 281).

Turner's frontier thesis held sway from the late 1890s to the 1930s, during which time historians and the public accepted it as the key to understanding the nation's past. Meanwhile, in 1931 Walter Prescott Webb put forward his well-known thesis that dry land shaped the character of Euro-Americans in the American West. The 20-inch rainfall line that runs through the middle of the United States, Webb pointed out in his book *The Great Plains*, separates the humid East from the dry West. In the humid East, Webb argued, America's institutions were based on water, timber, and land that was often rolling to mountainous. West of the 20-inch rainfall line in the Great Plains, water and timber were withdrawn, and the land was now flat. Webb's thesis stated that, when Americans moved west of the 20-inch rainfall line, their institutions broke down and had to be modified. He gave as examples barbed wire and sod houses used in response to treelessness, techniques of dry farming and wells drilled to tap groundwater employed to overcome semiaridity, and giant wheat combines and fast railroad locomotives used to conquer the vast flatness (Webb 1931, 3–9 ff.). West of the Great Plains, Webb noted in later writings, lay the eight "desert" states, the "heart of the American West," where conditions were so dry that in order to live Mormons and others had to cluster at oases and irrigate crops (Webb 1957, 26, 28, 31).

Turner's free land and Webb's dry land theses provided Americans with powerful yet simple interpretations of their national character and the character of people in the West. Each thesis seemed to be as solid as the land on which it stood. But each soon had detractors. Available land on the western frontier, Turner himself eventually recognized, did not serve as a "safety valve" for factory workers in the East because such workers lacked the resources to move to a new frontier. Thus, the frontier did not promote social democracy. And the ideas that frontiersmen were freer, more individualistic, and more nationalistic than were Easterners were also found to be overstatements (Billington 1973, 457, 459–65). Meanwhile, critics found defective Webb's thesis that the 20-inch rainfall line represented a cultural fault along which pioneer farmers stalled to make adjustments. In Nebraska and Kansas in the 1870s and 1880s, farmers systematically advanced well beyond this rainfall line into the Great Plains, especially along river valleys; and along the 98th meridian in Texas, where Webb spent his formative years at Ranger, Kiowas and Comanches, not lower precipitation, deterred settlement from 1860 to 1875 (Wishart 1987, 258–59, 271). To be fair to Turner and Webb, both theses to this day have merit, yet under scrutiny each seems simplistic. And each is most decidedly Eurocentric: Turner's advancing frontier and Webb's West were not, after all, empty of people.

We think that a third land-based thesis—namely, that of the *homeland*—

offers another holistic interpretation of the American character and the American West. The United States is a nation of immigrants. Native Americans, the first immigrants, were followed many centuries later by Euro-Americans and African Americans, and the influx continues today with Latin Americans and Asians. These immigrants and their descendants assimilated at varying rates and to different degrees into one American society, and for them the United States is, of course, their homeland. Where they settled these immigrants also developed some kind of attachment to place, some kind of individual homeland. To be sure, along the way a mainstream westward movement displaced Indian nations and overran early Spanish and French folk societies. Yet underlying today's pluralistic American society are individual homelands—large and small, strong and weak—that endure in some way. Our argument, then, is that within the single American homeland a number of lesser homelands capture the outcome of immigrant colonization. The mosaic of these homelands to which people bonded in greater or lesser degrees, we argue, affirms in a holistic way America's diversity, its pluralistic society.

The Analysis

In this volume geographers in commissioned essays discuss 14 American homelands. The areal coverages of these homelands are shown in a general way in map I.1, where we sort the homelands into those that are ethnic (eight) and those that are self-conscious (six). This anthology does not purport to include all possible homelands in the United States. The list of contenders would include the Sioux and many other Native American groups, places with clear outer limits such as the Sand Hills in Nebraska and the Willamette Valley in Oregon, as well as Hawaii and Puerto Rico. Nor does the anthology discuss the dozens of small ethnic islands found, for example, in the Upper Midwest— pockets of Germans, Scandinavians, Czechs, Luxembourgers, and others— because we are not convinced that they constitute authentic homelands. We think that one community of Swedes in Lindsborg, Kansas, or one urban barrio of Mexican Americans in Fresno, California, constitutes an island or ethnic enclave, not a homeland. These exclusions raise the question of what exactly a homeland is.

The concept of a homeland is not new but only recently has it been used by cultural geographers. In 1971, Alvar W. Carlson wrote a dissertation about

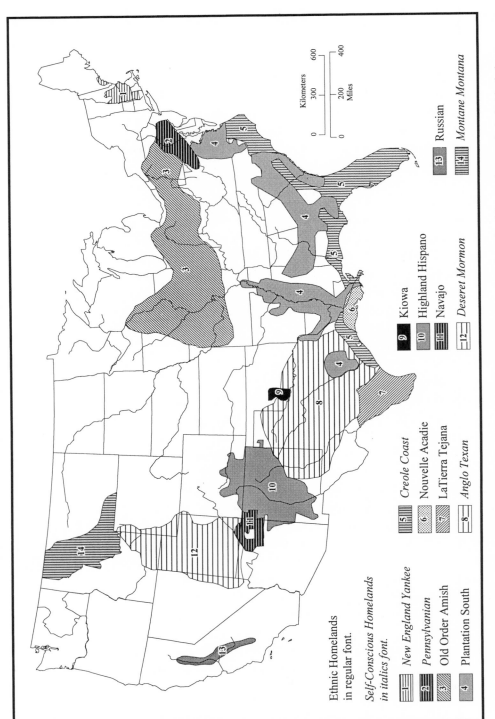

Ethnic Homelands
in regular font.

Self-Conscious Homelands
in italics font.

▥ 1	*New England Yankee*
▤ 2	*Pennsylvanian*
▨ 3	Old Order Amish
▦ 4	Plantation South

▥ 5	Creole Coast
▦ 6	Nouvelle Acadie
▧ 7	La Tierra Tejana
▦ 8	*Anglo Texan*

■ 9	Kiowa
▦ 10	Highland Hispano
▦ 11	Navajo
▦ 12	*Deseret Mormon*

▦ 13	Russian
▦ 14	*Montane Montana*

Map I.1. Homelands in the United States, 2000. Several homelands shown overlap; readers should consult maps in individual chapters (as numbered on this map) for clarification.

the Spanish-American homeland in New Mexico's Rio Arriba. Although he did not elaborate on the precise meaning of a homeland, Carlson (1971) did suggest that the foundation of a homeland rests on a people's ability to acquire, use, and retain land. Nearly two decades later, when synthesizing his many re-search efforts on the Rio Arriba, Carlson (1990) again used the term *homeland* in his *Spanish-American Homeland: Four Centuries in New Mexico's Río Arriba*, yet again he left the job of defining the term to others. Writing about the same part of the world, Richard L. Nostrand became one of those others. In *The Hispano Homeland* (1992, submitted to the University of Oklahoma Press in August 1990), Nostrand went beyond Carlson in two ways: Areally, he mapped and discussed the entire Spanish-American subculture including Carlson's Rio Arriba, and conceptually he took a stab at a geographical definition of a home-land: "The concept of a 'homeland,' although abstract and elusive, has at least three basic elements: a people, a place, and identity with place. The people must have lived in a place long enough to have adjusted to its natural envi-ronment and to have left their impress in the form of a cultural landscape. And from their interactions with the natural and cultural totality of the place they must have developed an identity with the land—emotional feelings of attach-ment, desires to possess, even compulsions to defend" (1992, 214).

Nostrand's interest in homelands piqued the curiosity of Lawrence E. Es-taville. During the 1990s, the newly created American Ethnic Geography Spe-cialty Group, under Estaville's urging and leadership, organized a series of pa-per and panel sessions to discuss the concept of the homeland. The first occurred at the Association of American Geographers' meeting in Miami in 1991, and they continued in San Diego (two sessions) in 1992, Atlanta in 1993, San Francisco in 1994, and Boston in 1998. Geographers who were brought into the discussions—including many of the authors of chapters in this vol-ume—made it clear that Nostrand's people, place, and identity with place needed broadening. People and place could stand, but bonding with place im-proved on identity with place, and control of place, also time, needed to be added. These five homeland "ingredients" became the underpinning of essays on homelands in the United States published as a special issue in 1993 of Al-var Carlson's *Journal of Cultural Geography*, guest edited by Nostrand and Es-taville.

In the 1990s, as our discussions evolved, Michael P. Conzen voiced some dissenting opinions. He thought that the five components of a homeland proposed by Nostrand and Estaville—people, place, bonding with place, control of place, and time—lacked more specifically defined underpinnings. He also questioned whether, aside from certain Native American or mestizo groups, homelands existed at all in the United States, given the recency of its immigrant population. For the special issue of the *Journal of Cultural Geography*, Conzen (1993) accepted our invitation to voice his views. He again accepted our invitation to express his counterpoints in this volume's final chapter. We applaud Conzen for crafting criteria that are conceptually more specific than our own, and we welcome his probing questions as to the authenticity of American homelands. But we respectfully disagree with him on two points: Conzen's nine criteria laid out under the headings identity, territoriality, and loyalty, we think, are too numerous and cumbersome to be easily remembered and readily used. More to the point, Conzen draws a strong parallel between homelands and nation-states, an association that may fit a European model of homelands, in which more homogeneous populations bound together by nationalism formed single political units in the last two centuries, but his nation-state linkage does not fit America's multiple immigrant peoples and their homelands. In the United States, unlike Europe, homelanders have not sought political autonomy.

And so once again we employ our five simple homeland components: *a people, place, bonding with place, control of place*, and *time*. We think that these five criteria define the American homelands in this volume. We hope that readers will be able to subsume Conzen's nine criteria under our five. For example, Conzen's discussion of territoriality with its three criteria—control of land and resources, dedicated political institutions, and a coherently manageable spatial unit—fit under what we mean by control of place. We also hope that readers will understand our attempt to separate clearly our use of the term *homeland* from *nationalism*.

The first component, *a people*, is self-evident. Resident populations having homelands in the United States are either ethnic groups, which in this volume include the Old Order Amish, African Americans, Cajuns, Texas Mexicans, Navajos, Kiowas, Highland Hispanos, and Russians; or they are groups that are self-consciously aware of their differences, which in this book include New England Yankees, Pennsylvanians, the Romano-Caribbean peoples of the Creole Coast, Anglo Texans, Mormons, and Montane Montanans. These 14 peoples once possessed folk cultures—meaning that they had basically self-sufficient economies, little occupational specialization, weakly developed individualism, and order maintained through family and church—but in the United States folk cultures have largely been replaced by American popular culture. We are hard pressed to suggest a minimum number of people required for a homeland to exist. Steven M. Schnell calculates that Kiowas, our smallest group, have only 8,600 members, half of whom live in Schnell's study area. And it does not seem necessary for a people to constitute a majority population in their homeland. The Old Order Amish, Ary J. Lamme III observes, are a small minority even in Ohio, where they are most numerous. It does seem necessary, however, that the homelanders themselves have a level of recognition that their homeland exists.

The second component, *place* or territory, is equally self-evident. The homelands discussed here range in size from several counties for the Kiowa in Oklahoma to several states for one-time Yankee New Englanders, or African Americans, as shown by Schnell, Martyn J. Bowden, and Charles S. Aiken, respectively. To label the entire United States as the homeland of Americans, as Conzen does, we think is to confuse homelands in their North American context with the concept of the nation-state, which is a higher level of political identity for a people who have established full sovereignty. Most of the 14 homelands discussed here are contiguous, yet this is certainly not true for the Old Order Amish and African Americans in the Plantation South. Some of the 14 homelands are clearly functional regions with interaction between central nodes and peripheries. These linkages are perhaps best exemplified by Salt Lake City and its Temple Square and Latter-day Saints (LDS) headquarters as the functioning node of Mormon Deseret, as explained by Lowell C. Bennion. With the major exceptions of the Plantation South, the Creole Coast, Cajuns, Old Order Amish, and Pennsylvanians, our homelands do not directly overlap. With the exception of Navajos, who constitute nearly 100 percent of their homeland's population, the ethnic homelands discussed have been overrun by mainstream society, diluting and thus weakening their homelanders' connections. Estaville discusses how Anglos so intruded into the Cajun homeland.

The key homeland component is *bonding with place*. This tie happens when a people adjust to the natural environment, stamp that environment with their cultural impress, and from both the natural environment and the cultural landscape create a deep sense of place. Yankee New Englanders, explains Bowden, adjusted to severe winters in the 1600s when building "large" houses—the essence of Yankee identity—to include massive central chimneys and additional exterior insulation in clapboards and shingles. Sense of place is an elusive concept. For the Highland Hispanos, Nostrand writes, it means identity with the *patria chica*, the village and its surrounding land; Hispanos seem to know intimately every bump on the land and every curve in the road. For the Pennsylvania homelanders, argues Richard Pillsbury, identity with place has less to do with a sense of community that sprang from the ethnic blending of primarily English, German, and Scotch Irish peoples than with an attachment to the distinctive towns and farmsteads fashioned through an amalgamation of ethnic influences.

Landmarks have great symbolic importance in creating a sense of place. For Kiowas, emphasizes Schnell, Rainy Mountain has such symbolic significance. Four mountain peaks, also four colors, positioned in each cardinal direction, Stephen C. Jett explains, define the spiritual ground of the Navajo homeland. Shrines such as the Alamo, Terry G. Jordan-Bychkov suggests, have sense-of-place significance to Anglo Texans. Bonding with place thus means that a people shape an area with their culture, and the area in turn shapes them: Feelings of attachment and belonging develop. If threatened, desires to possess become compulsions to defend, as happened in 1857–58 in Mormon Deseret. Of the homelands discussed, the Navajos', whose sacred ground is so heavily layered with memories and significance, certainly nears the pinnacle of place bonding.

A fourth component, *control of place*, facilitates bonding with land. Control is often achieved by owning land. For the Old Order Amish, Lamme finds that land ownership is mandatory. What lures an Amish church district to a peripheral location is the availability of contiguous blocks of farmland that are affordable because they are marginal. What preserves Kiowa Country as a homeland, underscores Schnell, are the 160-acre allotments still in the possession of tribespeople. Increasing home ownership by African Americans in the Plantation South, notes Aiken, signifies commitment to those areas as home.

Control is also achieved politically. Daniel D. Arreola shows how Texas Mexicans in their Tierra Tejana successfully converted their demographic plurality into political control at the ballot box. Indeed, emphasizes Jordan-Bychkov, the threat to Anglo Texans posed by ever growing Texas Mexicans beyond La Tierra Tejana nourishes and strengthens the Anglo Texans' sense of homeland. What defines the emerging homeland in western Montana, argues John B. Wright, is the philosophical struggle played out politically between developers and conservationists. Loss of political control can also bring on the unraveling of a homeland concept. The loss of the Civil War in 1865, reports Jordan-Bychkov in his second essay in this volume, triggered the demise of political control and the erosion of the homeland concept held by the Romano-Caribbean people of their Creole Coast. And the demise of the French language among Cajuns, Estaville stresses, weakened all aspects of Cajun control of their homeland.

Finally, bonding with land takes *time*, the last component. There is no set answer to how much time is needed. The Old Order Amish, Lamme argues, when peopling peripheral homeland areas, perhaps because of their closeness to land, develop strong bonds within a generation or two. Despite their recency, Slavic Russians in California's Central Valley constitute an "emerging" homeland, according to Susan W. Hardwick. But the strongest bonds between people and place require several centuries to develop. Some would argue that in Europe "real" homelands (that are not nation-states) are places like Scotland, Bavaria, Normandy, and Catalonia. But this view overlooks the Navajo who, emphasizes Jett, with the exception of four years (1864–68), when they were forcibly removed to Fort Sumner and the Bosque Redondo Reservation, have lived in the same place since the 1400s. It also overlooks the Highland Hispanos who, writes Nostrand, have been in place except for the time after the Pueblo Revolt (1680–93) since 1598. Indeed, homelands that have perished, like Bowden's Yankee New England or Jordan-Bychkov's Creole Coast, leave a "residue of regionality," as Jordan-Bychkov puts it.

The Outcome

So, drawing on our 14 examples, what can be said about homelands as a geographical concept? Homelands are, we suggest, specific places to which ethnic or self-consciously different peoples have bonded emotionally—with the aid of their control of place through time. Bonding with place, the key element, is strongest among ethnic groups whose folk cultures have been in decline. We would categorize homelands as special kinds of cultural regions, but they go beyond the delimitation of the multiple traits of a single cultural group. A group's relation to place, which is the key element in the concept of a homeland, is not central to the concept of a cultural region. Because people-place bonding is a process rooted in the human-environment, or man-land, tradition of geography, we find that the homeland concept has greater affinity to cultural ecology than to cultural regions. Homelands are, in the words of Jordan-Bychkov, "special culture areas."

As special culture areas where a group's relation to place is the key, homelands are not homogeneous or monolithic areal entities. This variance is because a group's relation to place differs in intensity within the region delimited as the homeland. In Nostrand's Highland-Hispano Homeland, Hispanos have strongest attachment to their local *patria chica*, literally the small fatherland, meaning the village and its surrounding land, and regionally the degree to which Hispanos have a sense of place decreases with declining Hispano percentages from the homeland's core to its periphery. In Jett's Navajo homeland, identity exists at several levels: for the Navajo country as a whole, for community bands and chapters, and for extended families and clans within communities. In both homelands, the smaller the level areally, the stronger the identity. And in Bennion's Mormon Deseret, attachment to the locale where individual Mormons "feel most at home" surpasses attachment to the homeland as a whole. Variations in the intensity of a people's sense of place are, of course, difficult if not impossible to measure. Nevertheless, in trying to understand homelands as a geographical concept, the existence of gradations means that homelands are not uniform regions.

Our aim in discussing the parameters of American homelands is not to introduce new jargon to geography but to try to understand and present a concept that is intrinsically geographical and that for many evokes powerful emotional feelings about place. And our aim is also to offer a thesis that we believe has merit in helping to explain the American character. We recognize that for mainstream American society, values of popular culture usually place little emphasis on attachment to place: When given the opportunity to retire elsewhere, many Americans move to Florida, Arizona, or some other part of the Sunbelt; how much they miss their original place is uncertain. But then there is a more traditional segment of American society represented somewhat imperfectly by our 14 homelands. For these more traditional peoples, affection for place and attachment to place are not shallow: Hispanos who have moved to California, African Americans who have gone north, and Kiowas who reside in Oklahoma City move back to their homelands when they retire or even before. Their identity with place endures.

And so we argue that in a nation of immigrants with an ingrained cultural pluralism—two attributes that differentiate the United States from many of the world's other nation-states—homelands have shaped an American sense of place. The aggregate of the thinking of traditional peoples about place, we believe, has only strengthened national solidarity and has enhanced those feelings of loyalty that underlie that higher level of identity—nationalism. Homelands, then, account for those human values that are rooted in place:

· a love for one's birthplace and home;
· an emotional attachment to the land of one's people;
· a sense of belonging to a special area;
· a loyalty that is defined by geographical parameters;
· a strength that comes from territoriality;
· a feeling of wholeness and restoration when returning to one's homeland.

And so if free land to some degree shaped the American character, and if dry land to an even greater degree shaped the character of people in the American West, then homelands to some degree shaped the American sense of place. They help to explain the American cultural mosaic. That America is a land of many peoples and many homelands seems complex and benign when compared to the simplicity and power of free land and dry land as explanatory theses. Yet what underlies the American character in an important way is the aggregate of the thinking of the traditional peoples whose sense of homeland has been strong. The 14 homelands that follow are the evidence.

HOMELANDS

CHAPTER ONE

The New England Yankee Homeland

. . . .

Martyn J. Bowden

H. L. Mencken found the term *Yankee*, first recorded in 1758, to be a derisive Dutch expression directed at English colonial yokels in Connecticut. In the 1760s *Yankee* spread rapidly to encompass plebeian New Englanders, and by the time of the Revolution it meant New England American and anti-British patriciate—and "Yankees began to take pride in it" (Mencken 1936, 111). *Yankee* does seem appropriate as a term for the first European-derived indigenous American culture in New England and for the new home-grown plebeian culture and social order that diffused, with varying degrees of receptivity, throughout most of settled New England by the eve of the Revolution. The drawback is that applying it to New England before the 1760s is anachronistic.

In this chapter I try to pinpoint in time and place the inception and development of this Yankee culture and a Yankee homeland. I find that an American Yankee culture evolves primarily among East Anglians, but with measurable input from West Country and Southeastern Englishmen, in a lowland hearth adjacent to Boston between approximately 1645 and 1680. I use settlement patterns and vernacular houses in my attempt to define the attributes of this new Yankee culture. I then delimit a "pure" Yankee homeland found in upland areas beyond Massachusetts Bay between 1680 and 1790. Regions where Yankee culture diffused outward in the eighteenth and nineteenth centuries, invariably in diluted form, to be superimposed on observably different regional cultures that developed in the seventeenth century, I suggest, cannot be thought of as Yankee homeland.

These areas are termed here the Yankee periphery, as are frontier areas settled after 1770 where a sizable minority of settlers came from the Yankee periphery. The homeland itself, I find, experiences dilution, decline, and eventual demise from the center outward at least from 1790 onward, yet clear evidence can be offered why a Yankee homeland once existed.

English Source Areas in the Seventeenth Century

Using genealogical evidence and place name transfers, I mapped for the seventeenth century the counties in England that supplied emigrants and the destinations of these Englishmen in New England (Bowden 1994a). The genealogical data were originally gathered by Charles E. Banks (1937), and from his compilation I could map 2,451 individuals in England and New England from 1620 to 1650 (Bowden 1994a, 76). Place names transferred from England to New England were generally those of preindustrial market towns and villages familiar to the settlers, which I mapped for the period 1620–1720. Considerable agreement existed between the two sets of data for 13 of 15 cultural beachheads found in New England; further study is needed to sort out discrepancies in coastal New Hampshire and the Connecticut Valley of Massachusetts. In England both data sets also show a clear hierarchy in five regional emigrant source areas.

Of the five regional source areas in England in the seventeenth century, East Anglia was clearly the leader in population numbers (map 1.1). Suffolk, Essex, and Norfolk, all counties in East Anglia, led all other counties in number of emigrants (1620–50) and in transfer place names (1620–1720) (Bowden 1994a, 74, 110). The Stour Valley on the Suffolk-Essex border was an especially important emigrant source. In decreasing order after East Anglia came the Southeast (notably Kent and Hertfordshire) and London; the West Country (especially Devon, Somerset, Dorset, and Wiltshire); the North (Yorkshire); and the Midlands.

East Anglia from Domesday (1086) onward was England's most populous area. It had a strong legacy of Danish-based freeholding (Postgate 1973, 306–8). This had contributed to land consolidation and enclosure in wood-pasture East Anglia, whence came the majority of East Anglian settlers in Massachu-

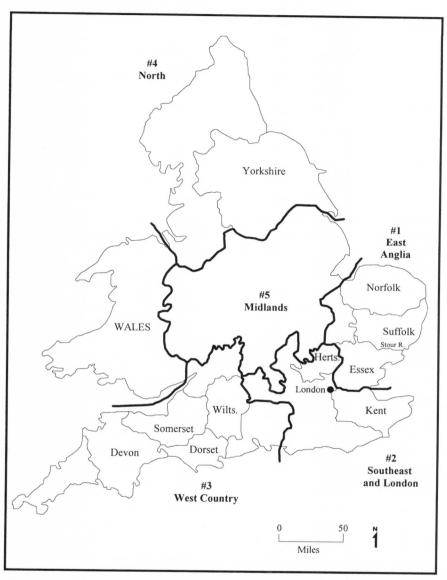

Map 1.1. English source areas for emigrants to New England in the seventeenth century. Shown are five regions (ranked by importance) and selected counties. The map is based on emigrant numbers (1620–50) and transfer place names (1620–1720). *Source:* Bowden 1994a, figs. 1, 3, 12.

setts Bay. The tendency toward enclosure was reinforced by the demands of the market—namely, East Anglia's strong medieval trade connections with nearby London and with the technologically advanced Low Countries, and its position as the preeminent industrial region in England from the fourteenth to the eighteenth century (Postgate 1973, 284–85; Roxby 1928, 156–58; Thirsk 1967, 48–49). The result was a landscape of enclosed farms with large closes "divided for practical convenience between several crops," shut in by large hedges with "their chequer board pattern broken at intervals by extensive woodlands" and a settlement pattern of isolated farmsteads, occasional hamlets, and small market towns (Thirsk 1967, 45–46).

This deeply ingrained land system and settlement pattern of wood-pasture individualism dominated the landscape of the Puritan diaspora in Massachusetts Bay in the early 1630s (pre-Yankee phase), for example, in Watertown, "a Plantation for husbandmen principally," which was laid out in a dispersed settlement pattern of compact farmsteads with common herding practices "but no hint of common arable field regulations" (Allen 1982a, 128). But by the middle 1630s the Puritan oligarchy, consisting of Governor Winthrop, the General Court, and the university-educated Puritan orthodox clergy, had set their goals to implement formal cultural change. They saw the Watertown model as antithetical to their formal plans for community formation (Wood 1978, 39–40, 67–68; Powell 1965, 92–95). They preferred the Boston-Sudbury model with its annular nucleated village (Rutman 1965, 36–39; Powell 1965, 6–7, 74–75, 178–79) common in the mixed farming of the more traditional, manorialized, East Anglian areas of the north and east where some consolidation of commonly held arable farm strips had occurred, but where cooperative husbandry was still practiced (Thirsk 1967, 40–46; Postgate 1973, 322–24).

Accordingly, the General Court in 1635 ordered that "hereafter, noe dwelling howse shalbe halfe a myle from the meeting house, and in any plantacon, graunted att this Court, or hereafter to be graunted without leave from the Court." This is supported by an anonymous "Essay on the Ordering of Towns," written about 1635 and found in the Winthrop papers, which essentially prescribes the annular nucleated model (Wood 1978, 39–42).

There were two effects of the Puritan Order in the proto-Yankee phase. First, individuals in established towns, for example, Watertown and Hingham, resolved to settle together compactly in townhouse plots, but they did nothing to implement their resolutions. Second, in several primary towns founded in the late 1630s compact village settlement occurred, but these were (with one exception) exposed frontier towns 25 miles or more inland from Boston, and although each was largely settled by people from mixed farming regions where nucleated villages dominated, they soon began to disperse in the 1640s (Wood 1978, 59–68).

Faced with an ancient individualism reinforced increasingly by a Boston market-driven economic individualism, the Puritan elite repealed the Puritan Order in 1640. No more agricultural villages were laid out in the Boston hinterland. What happened to the few compact villages is demonstrated in Sudbury. Agricultural village became hamlet near the meetinghouse on the common as farmers moved from village homelots to isolated farmsteads in the outlands (Powell 1965, 150–86; Wood 1978, 75–85).

The General Court's alternative to the Puritan Order was community formation via the hiving-off of new towns from the overly large primary towns (Wood 1978, 168–78, particularly map 177). This process of "legislative surgery," contingent upon the building of the central meetinghouse, the sitting of a Congregationalist (Harvard-educated) minister, and the organization of a covenanted congregation "preserved and renewed the communal ideal which was tied to settlements of no more than several hundred families" (Brown 1978, 48; Lockridge 1970, 3–22, 93–118).

This compromise between informal and formal cultural processes (Foster 1960, 12–13) in Massachusetts Bay led to levels and types of dispersion not found in East Anglia: hamlet with meetinghouse in the earliest towns and isolated meetinghouse on the common in later towns, each surrounded by a sea of scattered farms that tended toward linearity along roads and rivers. These are the signatures of the formative Yankee cultural landscape of the 1660s not found elsewhere in New England in the seventeenth century but soon to spread widely thereafter through pre-Revolutionary New England.

New England Beachheads in the Seventeenth Century

When East Anglians and other English regional colonists arrived in New England, they found an environment already transformed by Native Americans. The Algonquins, in practicing a shifting agriculture and biannual burning, had cleared much of once-forested southern New England to plant corn, squash, and beans and to improve hunting and gathering. To deny this accomplishment and to glorify and exaggerate their own, as well as to justify their dispossession and displacement of the Indians, the dominant Puritan elite in Massachusetts Bay in the mid-seventeenth century invented three myths: the Native Americans were ignoble savages, New England was a pristine desert wilderness, and the Puritan saints as God's sole helpmates created a Second England in 1630–50 (Bowden 1992a: 188–89; Bowden 1992b, 5–10). This imagined past erased memory of the preceding native Algonquin homeland.

What replaced the Native Americans were 13 confirmed English subcultural beachheads, each populated by significant numbers of settlers transplanted from four of the five English subcultural regions in particular (map 1.2). These beachheads were formed mainly during the Great Migration (1630–41). There were six West Country beachheads:

1. Coastal Maine (and the Piscataqua)
2. The lower Merrimack Valley with frontier extensions to the Concord Valley
3. Cape Ann (Gloucester)
4. Western Plymouth Colony (Plymouth, Bristol, Barnstable, and Newport counties)
5. Outer Cape Cod and the islands
6. The central Connecticut Valley.

The East Anglians had three beachheads:

7. The Massachusetts North Shore (Essex County)
8. The heart of Massachusetts Bay (Suffolk and Norfolk counties)
9. The lower Connecticut Valley (Hartford), a district settled during the 1630s diaspora from Massachusetts Bay.

People from London and the Southeast dominated three regions:

10. The entire southern New England coast (including the New Haven Colony) from Greenwich (Rhode Island) to Greenwich (Connecticut)
11. Middlesex County, Massachusetts Bay
12. The Kentish Shore of eastern Plymouth Colony and inner Cape Cod.

The one northern beachhead was the Rowley-Yorkshire enclave (13) on the Massachusetts North Shore. Beachheads in the Connecticut Valley of Massachusetts and on the New Hampshire coast appear from the genealogical record to be mainly East Anglian (Bowden 1994a, 98, 136–41).

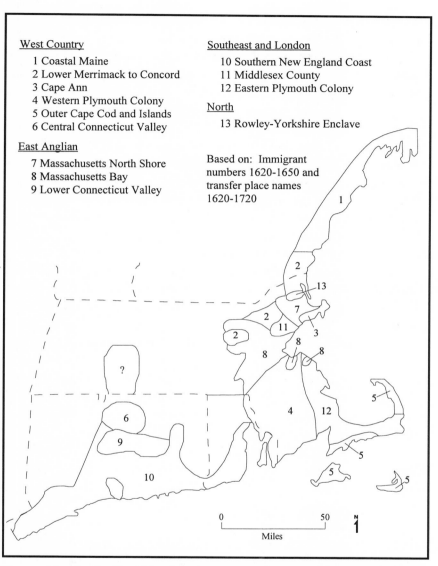

West Country

  1 Coastal Maine
  2 Lower Merrimack to Concord
  3 Cape Ann
  4 Western Plymouth Colony
  5 Outer Cape Cod and Islands
  6 Central Connecticut Valley

East Anglian

  7 Massachusetts North Shore
  8 Massachusetts Bay
  9 Lower Connecticut Valley

Southeast and London

  10 Southern New England Coast
  11 Middlesex County
  12 Eastern Plymouth Colony

North

  13 Rowley-Yorkshire Enclave

Based on: Immigrant
numbers 1620-1650 and
transfer place names
1620-1720

0   50

Miles

N

Map 1.2. New England beachheads in the seventeenth century. The 13 confirmed and two unconfirmed subregions are based on immigrant numbers (1620–50) and transfer place names (1620–1720). *Source:* Bowden 1994a, figs. 10, 28, 29.

Evidence presented by such scholars as Abbot Lowell Cummings (1979, 1994), Robert Blair St. George (1982), and David Grayson Allen (1982a, 1982b) increasingly points to the conclusion that particularly in the larger beachheads of seventeenth-century New England, the greater the number and the higher the proportion of settlers of the dominant English subcultural group (as suggested in place names and the genealogical record), the greater the probability that the construction techniques, building types, agricultural practices, settlement patterns, and field systems found there would be from the English region of origin of the dominant settler group. In the beachheads and in the entire settled area of New England the evidence also supports Foster's (1960, 10–20, 227–34) concepts of conquest culture and cultural crystallization.

The conquest culture was given shape by the culture type of the most numerous group of immigrants at least as far as informal transmitted elements are concerned. Informal processes of cultural selection ensured that, for example, dominantly West Country joiners' traits traveled successfully across the Atlantic to New England's West Country beachheads, but that many of the traits that were transferred failed to survive the stripping down or simplification process in the encounter with what Sauer (1941, 157–59) called the "lustier climate" of the New World. "Quick decisions, individual and collective, conscious and unconscious, had to be made . . . and the information on which settlers had to draw . . . was the knowledge that characterized their particular variants of [English] culture. The basic outlines of the new colonial cultures took shape at a rapid rate. Once they became comparatively well integrated and offered preliminary answers to the most pressing problems of settlers, their forms became more rigid; they may be said to have crystallized" (Foster 1960, 232–33).

The Lowland Yankee Cultural Hearth, 1645–80

The region in which Yankee culture crystallized between 1645 and 1680—the Yankee cultural hearth—is the immediate hinterland of Boston, as well as Charlestown and Salem in the adjoining coastal lowlands (map 1.3). I suggest that Yankee cultural formation in this Boston region emerged from three decisive forces, none of which was found as strongly elsewhere in New England: (1) East Anglian wood-pasture individualism (Thirsk 1967, 41); (2) the severe authority of the orthodox Puritan consociation of ministers and of the Puri-

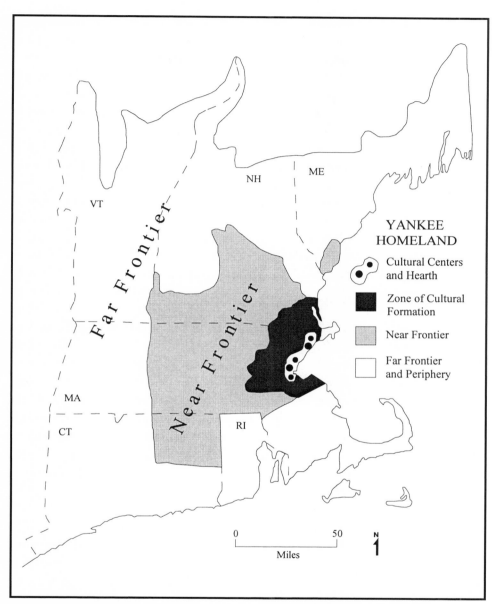

Map 1.3. The Yankee homeland in 1776. This homeland included the hearth (largest dot is Boston), the zone of cultural formation, and the relatively "pure" Yankee near frontier. In the Yankee far frontier Yankee migrants formed a majority in much of New Hampshire and Vermont. *Sources:* see text.

tan governor and General Court (Wertenbaker 1947); and (3) the protocapitalist values and success of the London-linked merchants in Boston (Rutman 1965; Bailyn 1964).

I also suggest that cultural formation fell into three phases: (1) pre-Yankee (early and middle 1630s), in which the informal culture of wood-pasture East Anglia dominated, with "each individual, through his pattern of living, [acting as] . . . a channel of cultural transmission to the contact area" (Foster 1960, 12); (2) proto-Yankee (late 1630s to middle 1650s), in which the dominant Puritan elite "groups in authority made the decisions they felt to be desirable, and attempted to enforce these decisions" (ibid.) in creative tension with Boston merchants and East Anglian farmers; and (3) formative Yankee (late 1650s to late 1670s), in which the success of Boston as the mercantile and imperial Little London of the Eastern Seaboard translated into a commercialization and comfortable agricultural prosperity for many farmers in Boston's immediate hinterland. That prosperity nurtured the cultural changes that completed the cultural transformation from East Anglian Puritan to Massachusetts Yankee.

Numerous, well-financed, well-supplied, and carefully prepared, the East Anglian Puritans led by John Winthrop immediately wrested away from all preceding groups in New England any initial advantage they may have enjoyed. Thereafter, they never lost their cultural dominance in New England. Strongly supported by a deeply religious and mainly East Anglian Puritan rank and file, the Puritan oligarchy consisting of Governor Winthrop, the General Court, and the university-educated Puritan orthodox clergy in the middle 1630s set goals to implement formal cultural change. The very presence of Winthrop in Boston and his perennial election as governor made his house the administrative-legal center of Massachusetts Bay and set in motion after 1634 a governmental and commercial advantage never since challenged in New England (Rutman 1965, 164–201; Price 1974).

Meanwhile, in the Boston Bay region, New England's only significant concentration of young, avant-garde joiners and housewrights from the technologically advanced areas of East Anglia and Kent developed a new housing vernacular as an architectural response to the demand of the only sizable group of modestly wealthy middle-class people who came to New England in the Great Migration (Cummings 1979, 25–32, 202–9; St. George 1982, 166). Consisting of five cumulated houseforms, this Yankee vernacular housing

Fig. 1.1. The New England "large," a five-over-five framed house, represents New England's highest form in vernacular architecture. Built in 1737, the massive central stone (gneiss) chimney of this house measures 15 feet on a side in the cellar and has five hearths (two 9 feet wide on the first floor). Between the double front door and the chimney stack is a triple-run staircase; single-run extensions reach the cellar and attic. Located in Sutton, Massachusetts, "Putnam House" is owned by Martyn and Margaret Bowden. Photograph by Richard L. Nostrand, 25 March 1998.

evolved in two phases—the pre-Yankee, to about 1640, and the true Yankee, to 1680—reaching its highest form in the central-chimney five-over-five New England "large" (fig. 1.1).

The evolution seems to have gone as follows: the English from East Anglia and Kent reproduced a hardwood box-frame, initially one-story but increasingly two-story house, with either a one-room houseprint with end-gable chimney or a two-room houseprint with axial chimney. The hardwood box-frame with studding is eastern English. The axial chimney, an avant-garde element recently diffused to England's southeast coastal ports from the Continent, was initially either East Anglian via Massachusetts Bay or possibly Kentish via the northern Pilgrim Colony. Shingling and clapboarding drew almost certainly on deep folk memory from the backwaters of Wealden Kent, made possible by abundant cedar on the South Shore. The combination of cedar clapboards and shingles on pine underboarding is a Massachusetts invention for colder winters made possible by abundant white pine. And the problem of storage in a hotter summer and colder winter necessitated abandonment of previous English practices and the invention of the cellar, probably in Massachusetts Bay but possibly in the Plymouth Colony (Cummings 1979, 25–32, 128–34, 202–9; Brunskill 1971, 106–7, 201–3; Brown 1979, 132–36; Clifton-Taylor 1962, 53–57).

The major breakthrough in the long-term solution to the problem of providing adequate and easily accessible storage and work space in a much colder winter climate than England's became common in the 1670s. It was the five-room houseprint and nearly square house of one or two stories with massive five-hearth chimney stack that distanced the heat source from exterior walls. The substantially uniform framing technology of oak and chestnut beams and joists bore distinct marks of a cultural fusion of the dominant East Anglian building tradition with Wessex traditions: notably a roof system of principal rafters and common purlins, a structural system of longitudinal summer beams downstairs and transverse summers upstairs, and of heavier timbering in general compared with East Anglian norms. The standardized triple-run staircase between chimney stack and front door with single-run extensions to attic and cellar are corollaries of this Massachusetts Bay adaptation to the new square chimney after 1670. The five-room plan consisted of two deep front rooms (hall and parlor with similar-sized chambers above), south-facing to maximize passive insolation, and a three-backroom plan centered on the new winter kitchen, flanked on the coldest northeast corner by the half-submerged dairy / buttery open to the cellar, and on the warmer northwest corner by the "little room" pantry. What Yankee folk took from London taste and Georgian canon was the centering of the axial stack and its alignment with the front door, symmetry in fenestration on front facade and end gables, and 12-over-12 or 8-over-12 sash: insufficient cause ever to label this great Yankee invention "Georgian" (Cummings 1979).

Thus, the one-story, one- or two-room, pre-Yankee houseprint with cellar grew to the so-called up-and-back with saltbox (integrated leanto) profile plan that triumphed in the last decades of the seventeenth century in eastern Massachusetts Bay (St. George 1982, 166), culminating, at first in the area northwest of Boston, in the highest form of the Yankee vernacular: the five-over-five New England large. A third variant was the two-over-five wide-based Cape Cod favored southwest of Boston. Combining these three standard houseforms with the one-room or two-room "American" cottage and the two-room, two-story hall and parlor I-house (modified English plans that dominated the first 40 years of settlement in coastal Massachusetts), the Yankee builder had five solutions to the problem of shelter. Furthermore, the developed Yankee plan allowed for expansion up and back by accretion from the one-room cottage via either I-house or Cape Cod to the saltbox and the New England large (Cummings 1979, 25–34, 50–84, 99–107, 128–34, 163–67, 202–9).

Perhaps the most obvious sign of change from English to American, from East Anglian to Yankee, and in the change of elites from Puritan to British patrician, are the place names on the land. As part of a strict political-cultural agenda of the orthodox Puritan elite and in strong support of the invented tradition of the "saints," the East Anglian Puritans of Massachusetts Bay were the first in the early 1630s to (1) exorcise Indian names (Shawmut, Agawam) accepted by the "Old Planters" who preceded them, (2) reject the generic-descriptive names given by the earliest Puritans (Trimountain, Newtown), and (3) view with disfavor religious names (Enon, changed to Wenham) and prospective names (Contentment, rejected for Dedham). Instead, they legislated the adoption of names of English (mainly East Anglian) market towns that proved (particularly on maps) that the orthodox Puritan saints, as God's helpmates, had created a new East Anglia, a Second England (Johnson 1654) in Massachusetts Bay. The cultural and political influence of Massachusetts Bay in the other beachheads is reflected directly in the timing and comprehensiveness of adoption in each beachhead of the Puritan orthodoxy in naming practices (Bowden 1994b).

Fifty years after establishing this practice of transferring English names, the new Americans of the Massachusetts Bay Yankee cultural hearth were the first to begin to cut back on the adoption of English transfer names, and on the use of the hitherto strongly favored East Anglian names. The flow of immigrants had dropped to a trickle after 1641 so that by the 1680s there was no longer any real folk memory of England. Furthermore, the English who had come over after the Restoration (1660) and following the Glorious Revolution (1689) were members of the "Court" in the British Empire's leading city in North America. New England merchants and the new Yankees of Boston's hinterland increasingly resented these representatives of England and those who emulated their culture and society. Instead of East Anglian names of towns from which the settlers of the Great Migration had come, the new places of the upland frontier of the Yankee hearth were given names (hitherto un-adopted by the General Court) of the remaining English Midland and North-ern counties and county towns (Worcester, Leicester, Rutland), whence few, if any, of the Massachusetts settlers had come, or they were given obscure names (Dracut) with no clear symbolic message. By the century's turn even these quasi-transfer names were giving way to simple names that used cardinal com-pass directions (Sutton, Weston, Easton), and, as a sign of resentments and up-risings to come, this frontier of cardinal points soon gave way to personal names of the British landed and governing elite in England and New England: royals (Lunenburg), Whig grandees (Walpole, Townsend), governors (Bellingham), generals, and to names of their mansions and palaces (Raynham). In each of the English beachheads of the seventeenth century, that beachhead becomes cul-turally Yankee roughly when and to the extent that alternative naming prac-tices begin to replace the use of English transfer names (Bowden 1994b).

Receptivity to the Yankee vernacular and to the five house solutions within the Yankee hearth itself during the eighteenth century differed greatly by beachhead subregion and sometimes by town. In the Charles River Basin in Massachusetts Bay, for example, the Yankee overlay came early and was thick on the ground. But in the remote areas of cultural backwash it was thinner and came in later, for example, on the East Anglian coast of the North Shore around Ipswich (few wide-based Cape Cod houses and a belated arrival of clas-sical five-over-fives), which had lost its mercantile competitiveness. Neigh-boring Rowley, that "drowsy corner" of the English North Country also on the North Shore (Allen 1982a, 19–54), and Gloucester, Rockport, and Cape Ann, of the English West Country, offered fewer examples of the Yankee ver-nacular than neighboring Topsfield, Newbury, and Ipswich (Cummings 1979). South of Boston the New England large and the saltbox were late to arrive on the border with the Plymouth Colony.

The Yankee Homeland Periphery

On the periphery of the Yankee homeland eight regions with subcultures evolved quite differently from the East Anglian heartland / Yankee hearth in the seventeenth century:

1. Coastal Maine
2. Coastal New Hampshire
3. Cape Ann
4. The (Old) Plymouth Colony and Lower Cape Cod
5. Outer Cape Cod and Martha's Vineyard–Nantucket
6. Rhode Island
7. The Connecticut Coast
8. The Connecticut Valley (map 1.4)

By cultural diffusion from the Yankee hearth and homeland, each of these areas became Yankee to varying degrees by 1800, yet none was so completely Yankee as to belong within the homeland. Salient attributes and benchmark events capture how these eight areas differed from the homeland.

West Country Mainers (region 1), religiously tolerant and hostile to East Anglian Puritans, living in posts for fishing, Indian trading and lumbering, and often transient, were a geographically fragmented, loosely ordered, centrifugal, materialistic, and uncommonly individualistic society that maintained strong connections with the English West Country (Hansen 1939, 78–80; Clark 1990, 18–50). They built elongated structures with end-gable chimneys, maintained their heavy timbering traditions in sawn log houses with four-sided overhangs, and transferred their vertical board construction technique (Candee 1976a, 15–44). Native American place names and descriptive names remained common and West Country transfer names (Portland, Bath) were not only common but were being used throughout the eighteenth century (Bowden 1994a, 98–106, 113–16; Bowden 1994b). The culturally based and long-standing dislike of Maine for Massachusetts was eventually translated into political fact with Maine's statehood in 1820.

The fishing-agricultural communities of the Cape Ann / North Shore region and of New Hampshire's Piscataqua (also with lumbering) (regions 2 and 3) were likewise maritime West Country beachheads and hotbeds of dissent from Puritan orthodoxy (Donald A. Smith, personal communication, 2000; Van Deventer 1976, 2–39; Hansen 1939, 64–65, 77–80). "The fishermen of Gloucester and Marblehead, many of whom were not Puritans, scorned to give up the least iota of their individual rights" and refused to become covenanted communities (Brown 1978, 48). Continuing West Country influences were expressed in plank framing, heavy timbering, transverse ground-floor summer beams, and principal rafter / common purlin roofing (Cummings 1979, 55–

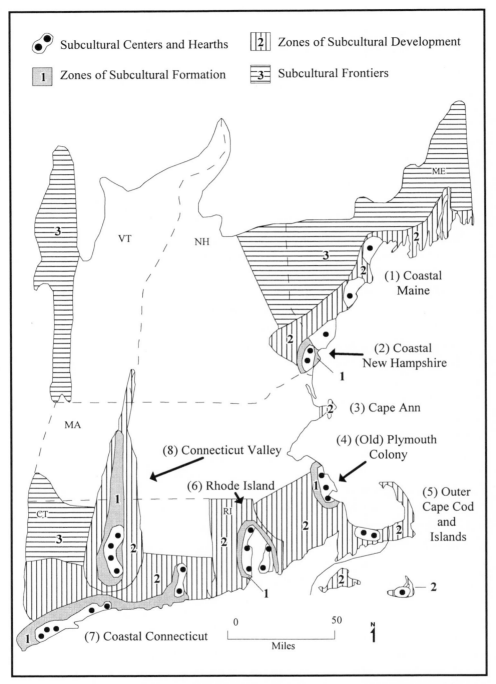

Map 1.4. Subcultures on the Yankee homeland periphery in 1776. Subcultural frontier areas lay beyond the Yankee homeland shown in map 1.3. *Sources:* see text.

62, 86–106; Cummings 1967; Candee 1976b; Candee 1992, 6–8). After spells of separation from Massachusetts, New Hampshire finally affirmed its individuality when it separated politically from the Bay Colony in 1691 (Van Deventer 1976).

The Plymouth coast extending to lower Cape Cod (region 4) attracted the Pilgrims: Separatists appreciably more tolerant of other sects than were the Puritans. The migrants were initially from the Southeast but with northern leaders. But the rest of the Old Colony including outer Cape Cod and the islands was soon overwhelmed by maritime-agricultural people from the West Country (Hansen 1939, 66–71; Bowden 1994a, 80–93, 120–29, 137–39). The Plymouth coast quickly became a cattle supplier to the Boston region, a "simple society with a primitive economy and a static social system" (Morison 1960, 138–42, 197–208). Greene (1988, 8, 18–19) finds that Plymouth "marked by geographical mobility, a high degree of individualistic behavior, and relatively weak ties of community" was, at the end of the seventeenth century, more like the Chesapeake colonies than it was like Massachusetts Bay. By 1691 this "comfortable" cultural backwater without a cultural center had failed and was absorbed by the Bay Colony, but its initial cultural distinctiveness lingered for another century and is reflected in vernacular architecture: (1) two transplanted English regional forms (neither of them East Anglian), 1620–50s; (2) Anglo-American mixes of two types (East Anglian and West Country) with distinctive regional forms, 1650s–1720s; (3) Yankee vernacular, accepted only in part, in which people showed an overwhelming preference for the Cape Cod house (1700–1790) (Deetz 1979, 43–59; Demos 1971, 24–35; St. George 1982, 166–67; Connally 1960).

Outer Cape Cod and the two Islands (region 5) is epitomized by Nantucket where a largely egalitarian society became more so with the widespread conversion to Quakerism, a sect severely persecuted by the Massachusetts Bay Puritans. Nantucket, against the inhabitants' wishes, was taken from New York and given to Massachusetts Bay in 1692 just as the Yankee vernacular began to diffuse outward from the Yankee cultural hearth. A century later no Cape Cods and few if any New England larges had crossed the sound and no more than ten full saltboxes had done so (Lancaster 1972, 2–252).

Societal pressures for plain living, uniformity, and thrift, and a deep antipathy toward the Puritan / Yankee heartland turned the Quakers' gaze instead to the half-house building tradition of the Connecticut coast and Newport and to the end-chimney, stone-ender tradition their ancestors had carried from Rhode Island to New Bedford (Hansen 1939, 70). In Nantucket they simply melded three traditions and took what little they needed from the Yankees who were still turning away their representative to the General Court in 1770. Presbyterians, Baptists, and Quakers, many of them descendants of West Country dissenters hounded from the lower Merrimack Valley or from the Massachusetts margins of the Plymouth Colony were, at the time of the Revolution, still barely receptive to Yankee culture (Lancaster 1972; Lancaster 1979, xiii–xxii).

The greatest deviant from Puritan orthodoxy and the Yankee mainstream, however, was Rhode Island (region 6) (Hansen 1939, 71–76; Daniels 1983). Religious tolerance of Portuguese Jews, French Huguenots, and Irish Catholics, for instance, made that colony more like Pennsylvania than New England (Greene 1988, 45–46, 173), and Rhode Island's northern and western borders with Puritan orthodoxy represented a veritable cultural fault that intrepid Rhode Islanders crossed at some risk (Hansen 1939, 71–76, 84).

Rhode Island's severing of direct cultural connections with Massachusetts Bay meant that the pre-Yankee house brought in by those expelled from Massachusetts Bay in 1637 had evolved by the second half of the seventeenth century into the stone-ender, a distinctive Rhode Island house type with a projecting end chimney and a West Country plan: long and low slung, with doors that opened into their hearts, with vertical planked, up-braced framing, and principal rafter / common purlin roofing (Upton 1979, 18–33; St. George 1982, 166–67).

Coastal Connecticut (region 7) consisted of settlers from the Home Counties adjacent to London, many of whom came directly to the region and maintained maritime connections with the London region. Some of the compact "commercial villages in which agriculture predominated" (Wood 1978, 145–52) and which were unique to the Connecticut coast (most of them with names ending in "-ford") were united early in the short-lived New Haven Colony, which, among the eight peripheral cultural regions, was the most similar to Massachusetts Bay in its Puritan orthodoxy. The vernacular architecture of the region exhibited various seventeenth-century eastern English housing characteristics abandoned early in the Yankee hearth and revealed a number of Dutch influences (Cummings 1994, 192–226; Kelly 1963, 21–68).

The Connecticut Valley (region 8) was settled from Massachusetts Bay in the middle 1630s because of the reputed and actual fertility of the land (Hansen 1939, 93–95). It attracted moderate Puritans already bridling at the heavy hand of Puritan orthodoxy and the formal cultural processes of the Puritan oligarchy. In complete contrast to Massachusetts Bay, settlement patterns were of the open-field, street village type with landholdings in long lots (Wood 1978, 149–52; St. George 1985, 29–40; Garrison 1991, 8–27). And Cummings's (1994, 192–226) recent studies reveal diverse, mainly East Anglian housing styles dominating the Connecticut Valley region throughout the seventeenth century and not the Yankee style rapidly developing in Massachusetts Bay. The "River Gods"—the hereditary aristocrats—soon controlled most of the property, profited as middlemen in Seaboard and trans-Atlantic trade networks, and assumed the role of cultural brokers. They coercively manipulated the landscape, favored distinctive artifacts of regional consciousness, such as hewn and framed overhangs, gambrels, disproportionately large and elaborate doorways and gravestones, and chests without drawers, and thereby perpetuated seventeenth-century style in the Connecticut Valley into the second half of the eighteenth century (St. George 1985, 29–40; Miller 1983; Kelly 1963, 59–64).

The Near Frontier of the Yankee Homeland, 1680–1790

The five elements of the Yankee vernacular came together on the western flank of the Yankee hearth in the eastern lowlands of Massachusetts Bay (Wright 1934, 14–19, 38–41). They were well known to the population growing rapidly but pent up in the lowland hearth awaiting the political settlement that came eventually with the Treaty of Utrecht in 1713. Thereafter, the population "swarmed" westward bearing with it the five Yankee house "solutions."

An Indian imprint in the uplands during the early seventeenth century, also that made by scattered and largely abortive European forays, was largely obliterated by the regrowth of the forest by the time the settlers on this near frontier (Brooke 1989, 5–13) penetrated the eastern uplands in earnest. On these uplands is imprinted the purest expression of Yankee culture that exists anywhere in the New England landscape: The five house types of the new Yankee vernacular and the landscape of isolated farmsteads, each with separate barn and outbuildings, plus a "town center": "commonly a meeting house and burying ground around a meeting house lot" (Wood 1997, 2).

The result was a sharp cultural fault line in the landscape where the uplands met the lowlands. The pure and simple Yankee vernacular of the uplands in the eighteenth century lay to the north and west (Steinitz 1986), with the complex landscapes of the lowlands to the east showing "relict features" from the Algonquin, East Anglian (pre-Yankee), and transitional (proto-Yankee) landscapes beneath the emergent formative Yankee landscape: the classic pattern of sequent occupance from the sixteenth through the eighteenth centuries in eastern New England (Whittlesey 1929; Ackerman 1941). And the Yankee homeland's cultural essence was in the zone settled by the first generation of Yankee frontiersmen after 1713: the "middle landscape" (Brooke 1989, xiii–vi) of the Worcester uplands: southeastern Worcester County, Massachusetts, and neighboring northeastern Windham County, Connecticut (Hansen 1939, 81–90).

To their upland home Yankees took their five house solutions and their lowland Yankee agricultural practices and settlement patterns. Beyond the ring of intensive agriculture (mixed husbandry, with emphasis on dairying and the cultivation of cereals) created by wood-pasture East Anglian farmers to supply Boston, there developed far-flung rings of extensive agriculture (cattle raising with some fattening on grain, and sheep farming beyond). Droving roads linked these rings to Boston. Meanwhile, in their newly formed towns, Yankee farmers lived in scattered farmsteads, each with clustered outbuildings. The town focal point, the meetinghouse, and its cluster of structures that included a school, blacksmith, and tavern, repeated the dispersed settlement pattern found in the lowland hearth.

After the 1740s, Yankees, predominantly from the near frontier, pushed north into still poorer, higher, colder, and recently dangerous Indian lands, for example, in northern Worcester County (map 1.3). In this second stage they were joined by Yankees from the cultural hearth pushing into New Hampshire west of the Merrimack. Later in the century, these Yankees of the eastern uplands vaulted the Connecticut Valley and settled the western uplands of Massachusetts and present-day Vermont and the northern uplands of New Hampshire and Maine. They were joined there by sizable minorities of settlers from the Yankee periphery (Hansen 1939, 76–104). The result, by 1790, was (1) a Yankee homeland consisting of the cultural hearth of eastern Massachusetts Bay and the near frontier of the adjacent eastern uplands and (2) a Yankee periphery consisting of the eight regional subcultures and hearths subsequently

Yankeefied in varying degrees after 1690, and of the far frontier of settlers who hailed from both the Yankee homeland and the Yankee periphery.

Yankees in a National Culture after 1790

About 1790 some New Englanders ceased to exist as "Yankee." Yankee culture was born in Boston and its immediate hinterland in the middle of the seventeenth century. Here it was so seriously challenged by the high-style culture of the British patriciate a century later that the American Revolution resulted (Bushman 1970, 272–88; Bushman 1984, 345–83; Brown 1978, 66–101). And here within New England is where the national culture first takes hold after the Revolution. Americanization and de-Yankeefication spread outward from this traditional hearth of culture, which was "the hole in the doughnut" of Yankee culture by 1800. By 1830 most of the Yankee homeland and all of the old Yankeefied subcultural regions of the Yankee periphery were part of a new convergent national culture that continued to exhibit artifacts and traits identifiable as Yankee. And by the late nineteenth century, even the "hill Yankees" of the old far frontier were more Americans and New Englanders than they were Yankees. The paradox is that the Yankee culture that existed in degrees around the far frontier became the center of a new "Yankee culture" with the word *Yankee* now redefined. Yankees are thrifty, shrewd, cantankerous, and eccentric: "Their wit is dry, and understatement is preferred to overstatement. . . . The Yankee is a Puritan soul and a Puritan Mind, but tinctured for three gradual centuries by the poison of worldly understanding" (Bearse 1971, 72). In the Housatonic Valley of the old far Frontier after the Civil War, the average Yankee "succumbed to the materialism that was sweeping the country . . . took his family to church, because it was . . . the thing to do. But his idea of God, let alone old Congregationalist doctrine, was extremely vague . . . like the rest of the world, he became an agreeable agnostic" (Smith 1946, 22–26, 433–37, 462–64).

After 1790 people from the settled lowlands and uplands of New England began to move in an expanding frontier to the north and west, carrying with them the second New England cultural landscape solution to the problems of settlement. Once again the "solution" originated in the lowlands. A great rebuilding based on imported English and increasingly Americanized styles was stimulated by capital generated in the mercantile ports during the post-

Revolutionary period and later by profits from industry generated by the water power of rivers crossing the lowlands. This adds a fourth landscape layer (after Algonquin, English immigrant, and Yankee) to the cultural complexity and sequent occupance of the lowlands. It makes impressive localized changes in the still legibly Yankee uplands, mainly in and around central places, and is the popular high style favored by the emergent elite in the lands settled by late Yankees between 1790 and 1830. This landscape of spired Bulfinch-style churches, central-hallway houses, and New England villages (Wood 1986, 54–63) is the archetypal landscape of the golden age of New England to many, particularly tourists. But in a region undergoing massive changes in the social order, already hell-bent on leading the nation into industry, and supporting the rise of its major city to the position of the nation's cultural capital, the cultural landscape of the fourth period becomes suddenly complex: more an expression of national (federal) culture than of the old Yankee regional culture.

The second and the last major wave of New England Extended (1830–60) is signified by the "new nonfolk, Greek Revival (national house) form typified by a door in the gable and one or two low wings off to the side. This temple-form house became the predominant type through the North and out into the Great Lakes area" (Glassie 1968, 129, 133). It existed as a second cultural layer in the region of the federal / republican frontier, a third cultural layer in the old Yankee homeland of the eastern uplands of New England, and as the fifth cultural layer in the coastal lowlands of the old Yankee hearth. Glassie reminds us that these expressions of popular culture (temple form and upright-and-wing) vastly outnumber the occasional folk elements of Yankee culture carried over from the New England vernacular of the Yankee homeland (1680–1790) and from the first (frontier) wave of New England Extended (1790–1830). The house forms remind us therefore that this final wave of New England Extended is an expression of American national culture and not of Yankee (vernacular) culture.

Conclusion

Thus, in a coastal lowland adjacent to Boston, a Yankee culture evolved among East Anglians especially between 1645 and 1680. Two expressions of this new Yankee culture were a dispersed settlement pattern with meetinghouse center and outlying farmsteads; and five vernacular house types that spoke to a Yankee ingenuity to adjust to severe winters through clapboards and shingles on underboarding and cellars as solutions for storage. In this Yankee lowland hearth these Yankee attributes were intermixed with the array of cultural baggage brought from England. But when taken to the uplands of Massachusetts, Connecticut, and southern New Hampshire between about 1680 and 1790, they stood as the pure and uncontaminated expression of a Yankee culture 35 years in the making in the hearth. From the Yankee homeland—both its lowland hearth and its more pure upland extension—Yankee culture diffused as a partial veneer to the eight coastal and riverine peripheral areas, and beyond them to the far frontier of New England and beyond to frontier areas referred to as New England extended. What challenged and undermined Yankee culture and a Yankee homeland was the spread of a national popular culture—an American culture—after about 1790.

The argument why this region of Yankee culture constitutes a Yankee homeland is clear: Yankees adjusted to their long severe-winter environment, they stamped that environment with their distinctive impress, and in the process they bonded with New England. Examples of adjustment include building houses with clapboards and shingles on underboarding, an innovation made possible by abundant white pine, and digging cellars for cold winter–warm summer storage, an invention made in Plymouth perhaps but probably in Massachusetts Bay. The Yankee landscape impress manifested itself in the five house solutions (one-story, one- or two-room cottage with cellar; the two-room, two-story hall and parlor I-house; the two-over-five room saltbox, as well as the Cape Cod; and the five-over-five room large) and in the settlement pattern of dispersed farmsteads with central meetinghouse marking the town center. These house types and settlement patterns to this day are the icons of Yankee culture and the important symbols of a Yankee's sense of place.

CHAPTER TWO

The Pennsylvanian Homeland

. . . .

Richard Pillsbury

Historical Development

Early Europeans viewed the lower Delaware Valley as a place to be explored, exploited, and conquered. Dutchmen settled at Fort Nassau south of Philadelphia in 1623, and Swedes arrived in 1638. Swedish colonists soon occupied scattered settlements from Tinicum Island (near Philadelphia) south to Fort Chistiana (Wilmington). The Dutch occupied the remainder of the Delaware River shoreline south to Lewes (map 2.1). This dual occupation lasted until 1655, when the Dutch captured the entire area. Few Swedes left after the change in ownership, and the English, in turn, acquired the lower Delaware in 1664.

The impact of this short occupation of the lower Delaware by Dutch and Swedish colonists is astounding. Neither group was numerically large: fewer than seven hundred Swedes and only a few thousand Dutch lived there in 1664. Yet their shaping of the landscape continued to be highly visible until recent times. Anglicization of that landscape began almost immediately in 1664. Peter Kalm noted in his journal "that before the English settled here they [the Swedish settlers] followed wholly the custom of Old Sweden; but after the English had been in the country for some time, the Swedes began to follow theirs [the English]" (Benson 1966, 273). Yet, Dutch- and Swedish-style houses and barns continued to be constructed for several generations and some relict structures remain.

English people dominated the lower Delaware politically and culturally after 1664. New Jersey was initially granted to Sir George Carteret and Lord Berkeley, but Berkeley soon sold his undivided share for £1,000 to Edward Byllinge, a Welsh Quaker. Byllinge appointed three Quaker trustees, including William Penn, to develop his American holdings. The trustees negotiated a division of the colony that gave Byllinge the western half, called West Jersey, by running a line diagonally across the colony from Little Egg Harbor in the southeast to a point near the current junction of the Pennsylvania, New York, and New Jersey borders in the north. A combination of English settlers from Long Island and southern New England and Dutch settlers from the lower Hudson Valley took up most of these lands, excluding Carteret's East Jersey from the future homeland at the very beginning. West Jersey was divided into a hundred geographically undesignated proprietaries that were offered to Quakers for immediate settlement. The first group arrived in 1675, settling near present-day Salem. The next two groups purchased adjacent proprietaries along High Street in Bridlington, later renamed Burlington. A group of Irish Quakers instituted the fourth settlement node in 1681–82 near Camden.

England largely ignored Delaware after its political absorption, allowing anarchy to rule the colony throughout the latter half of the seventeenth century. Growing tired of the complaints emanating from the colony, Charles II placed it under Penn's management soon after he established Pennsylvania. This arrangement pleased the residents, who looked forward to support in controlling both the pirates who raided the coast and Lord Baltimore's land agents who continually invaded their territory. But the colonists were doomed to disappointment and soon lobbied for the creation of a new charter to establish a separate elected legislative assembly and governor. Culturally, Delaware has never been strongly associated with any larger area, nor did it ever emerge as a distinctly individual place. Early Dutch and Swedish influences set the landscape pattern, but like West Jersey it was soon dominated by Englishmen. Even today it appears to be visually more akin to West Jersey than to Pennsylvania.

William Penn's enthusiasm for the West Jersey experiment quickly waned when it became obvious that only Quakers would be welcome in the new colony, even though the Society of Friends specifically preached religious tolerance. Penn approached Charles II and the Duke of York about obtaining

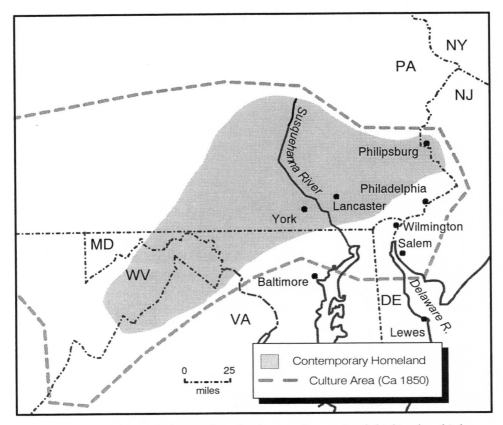

Map 2.1. The Pennsylvanian homeland, 2000. Conventional thinking has this homeland sprawling over rural southern New Jersey, eastern Pennsylvania, northern Maryland, and portions of Delaware, Virginia, and West Virginia. Once an accurate areal perception, over the past 50 years rapid suburbanization, regional economic restructuring, and the development of tourism have quietly reduced the extent of the region. The Pennsylvanian homeland, nonetheless, continues to be a viable entity within the eastern United States.

lands for an American colony in which real religious tolerance would be practiced. Owing Penn's family £16,000, Charles II agreed to settle the debt by granting Penn "a tract of land . . . lying north of Maryland, on the east bounded by the Delaware River, on the west limited as Maryland, and northward to extend as far as plantable" (Garber 1934, 65). These boundaries ignored the earlier Connecticut grant that extended from sea to sea, and it left the boundary with Maryland and Virginia to be settled after more was known about the western frontier.

Penn set about establishing a colony in 1681 by sending his cousin, William Markham, to Pennsylvania to begin making arrangements for the "Holy Experiment." Markham and three commissioners were instructed to locate and survey a seat of government before Penn's arrival in 1682. Searching along the Delaware River, the commissioners chose a site just north of its confluence with the Schuylkill. Thomas Holme, Penn's surveyor-general, laid out a rectilinear plan for the "greene Country Town."

Penn's open-door policy brought rapid settlement, and the commonwealth became the most ethnically diverse colony in America, if not in the world (Lemon 1972; Purvis 1987). Philadelphia, Chester, Wilmington, and even smaller Delaware River ports became centers for newly arriving immigrants who found unclaimed lands in Penn's Woods and began farming. The richness of the land, the diverse heritage of the early residents, and easy access to water transport within a few decades transformed Pennsylvania's subsistence economy to one of the most productive commercial agrarian economies in the colonies. The Delaware River formed the core of the newly developing region as settlement rapidly expanded westward to Blue Mountain and the Cumberland Valley and northward to the Delaware River water gap.

Cultural Foundations

In 1685, William Penn noted of Pennsylvania that "the People are a collection of diverse Nations in Europe: as French, Dutch, Germans, Sweeds [sic], Finns, Scotch, Irish and English; and of the last equal to all of the rest" (Myers 1912, 260). The colony met Penn's goal of being open to all, though in actuality three ethnic groups dominated in the eighteenth and early nineteenth centuries: English, various Germanic groups referred to loosely today as "Pennsylvania Dutch," and Scotch Irish.

The relative importance of each of these groups has been in dispute since colonial times. The Census Bureau attempted to lay the question to rest in 1909 with an analysis of the surnames listed on the original 1790 population enumeration schedules. Unfortunately, the Anglicization and ambiguity of many surnames precluded absolute ethnic identification of many, and the results were disputed almost immediately, though the strong dominance of the

English (59%) was resubstantiated in historical studies for decades thereafter (Purvis 1987). The American Council of Learned Societies revised these statistics in 1932, lowering the estimated percentage of English to 35, while raising the Germans to about one-third of the total. In 1980, McDonald and McDonald again lowered the English count to a mere 20 percent, but in 1987 Purvis's landmark study raised the English count to 25.8 percent. Purvis found it impossible to separate Scots from Scotch Irish and combined them (26.7%), making them more important than the English, whom he combined with the Welsh (25.4%).

Although revisionist history has reduced the count of English settlers, their importance outweighed their numbers. Though a numerical minority, Englishmen, because of their economic and political power, shaped much of the foundation for the commonwealth, including its basic legal code and language.

Settlement tended to cluster largely along ethnic lines in eighteenth-century Pennsylvania, although never absolutely. The difficulty of travel increased local isolation. Most counties had small "minority" settlements that seemed to prosper as distinct entities into the twentieth century. The English lived throughout the region but were most concentrated in the so-called English counties of Bucks, Montgomery, Delaware, and Chester, which surround Philadelphia (Lemon 1972). Scotch Irish settlers represented an important admixture in early-eighteenth-century Chester County, though they represented less than 20 percent by 1790. All things considered there seems to have been far more intermixing in all districts than is often suggested in the ethnic literature.

The Pennsylvania Dutch are the best known of Pennsylvania's non-British immigrants. These Germanic people emigrated to Pennsylvania from a variety of areas within central Europe, especially the southern Rhineland, Switzerland, and eastward. The term *Pennsylvania Dutch* stems from the Anglicization of *deutsch*. Many had only their German language and central European points of origin in common.

The first twelve "German" families arrived in October 1683 to settle on a 6,000-acre tract purchased by Francis Pastorious near present-day Germantown along the Schuylkill River. A flood of both sectarian and church Germans followed, including the Schwenkfelders in 1734 and the first group of Moravians in 1738. The majority of these people located in a band stretching from Frederick, Maryland, northeast beyond Philipsburg, New Jersey, with the heaviest concentrations found in Pennsylvania, including Lancaster County (71.9%), Northampton County (62.9%), western Montgomery County (56.6%), and Berks County (85.4%) (Purvis 1987).

More than 58,000 Germanic settlers from the Palatinate provinces entered Pennsylvania in the 28-year period after 1727. Most were pushed out of Europe for the same reasons—religious persecution and the collapse of the woolen industries following the Thirty Years' War. These "push" factors worked in concert with the extreme claims about the salubrious conditions in the New World emanating from an almost limitless volume of promotional literature widely distributed throughout Germanic Europe. Gottlieb Mittelberger, a German writer and traveler, became so incensed at the one-sided tenor of these pieces that he published a counter tract in 1754 in which he stated: "What really drove me to write this little book was the sad and miserable condition of those traveling from Germany to the New World. . . . For before I left Pennsylvania, when it became known that I wanted to return to Wurttemberg . . . numerous . . . begged me with tears and uplifted hands, even in the name of God, to publicize their misery and sorrow [while I was] in Germany" (Mittelberger 1960, 17). Most potential colonists ignored Mittelberger's plea and the migration continued unabated.

Popular history has tended to consider the Pennsylvania Dutch as culturally monolithic, whereas in fact a great deal of diversity in tenets and ways of life always existed. The variety of house and barn forms, as well as the many denominations of churches, is silent witness to the fact that these people were singular only when compared to non-Germanic settlers.

The Scotch Irish were Pennsylvania's third major ethnic group. Like the term *Pennsylvania Dutch*, this group's vernacular name is a misnomer. The term *Scotch Irish* is American and refers to those settlers of Scot ancestry who settled in northern Ireland after 1610 in reaction to efforts by James II to "Christianize" Ulster by reducing the dominance of Catholic residents. There is no European use of this term. Some ethnic historians argue that many migrants were fourth-generation residents of Ireland with little or no intact Scottish heritage. Known as Scotch Irish or Irish in America, these people were comparatively easily identified by their generally Presbyterian church affiliation and overall "Scot" physical appearance.

Moderate land rents and comparative freedom attracted about 200,000

Ulster Scots to America prior to 1717. Almost two million people left Ulster for North America after 1717. This great migration was triggered by the Woolens Act (1699), which brought depression to the Irish woolens industry; the Test Act (1703), which required all officeholders in Ireland take the sacrament of the Established Church; and rising rents based on an increasing value of the lands that Ulstermen leased from English landlords. Drought in 1717 and 1719 brought lower yields, but rents were not reduced commensurately, setting off the first major wave of emigration in 1717–18.

Migration ebbed after 1718, but the pattern was set. Another round of heavy emigration ensued each time economic and social conditions deteriorated. Word of the success of early migrants, the increasing importance of indenture as a means of paying passage to some colonies, and declining trans-Atlantic fares all favored a continued flow of migrants. Although almost every trans-Atlantic port in America received a share of these immigrants, the inexpensive lands and religious tolerance of Pennsylvania attracted the largest number to the Delaware River ports. The popular myth that the Scotch Irish fled the settled areas to concentrate primarily on the cheaper lands along the Pennsylvania frontier, however, is contradicted by the 1790 census. Almost half of the Scotch Irish lived in the largely German districts and the so-called English counties to the east (Purvis 1987).

The relative merits and qualities of each of these main groups, as well as those of the several dozen other nationalities, has been a topic of frequent discussion since before the American Revolution. In general, the Germans have been characterized as thrifty, hard-working, and excellent farmers. By contrast, the Scotch Irish are often described as shiftless, poor, and lazy farmers. While this question is only peripheral to the evolution of the Pennsylvanian homeland, it is a part of the larger question of separating the cultural origins of the various settler groups that form the culture as a whole. James Lemon, after an exhaustive examination of eighteenth-century southeastern Pennsylvania frontier sources, concluded that it seems "more sensible to approach Pennsylvanians of the eighteenth century as Americans with a western European background in which major differences in behavior and attitudes were the result of religious beliefs, social status, and economic circumstances, rather than attributable to a vague, elusive, and unchanging phenomenon called national character" (Lemon 1972, 227). Similarly, Ronald Clifton (1971) concluded that economic circumstances, not ethnicity, determined space allocation and by extension the house forms of residents of the region in the eighteenth century.

The national character question lends itself to a classic case of ambivalence in the most aggressive sense of the term. Those familiar with the area can see clear evidence of untarnished ethnic elements spreading throughout the region during the eighteenth and early nineteenth centuries. Simultaneously, it is possible to find landscapes dominated by elements that were blended from several ethnic sources. Ultimately one must conclude that while "pure" European examples can be found idiosyncratically across the landscape, most farmsteads, houses, barns, and other structures exhibit amalgamations of ethnic influences resulting from the American experience. This "blended" landscape existed by the late eighteenth century as Americans increasingly perceived themselves as separate from their European roots.

The dominance of an ethnic triad does not erase the continuing effect of the lesser groups who also settled the region in colonial times. Analyses of surnames in the 1790 census show a cultural milieu dominated by three groups yet interspersed with large numbers of residents of other nationalities. For example, Marshe (1801, 177) summarized the population of Lancaster, Pennsylvania, in 1744 as "chiefly High-Dutch, Scotch-Irish, some few English and unbelieving Israelites."

Identifying Characteristics of the Region

The visual landscape is this homeland's most easily identifiable feature. Differing little physiographically from surrounding areas, the area's cultural imprint is both bold and unique (fig. 2.1). The Pennsylvanian farmstead with its dominating two-story barn, two-story masonry residence, and assortment of outbuildings is the single most important regional landscape element, though the Pennsylvanian town too is a diagnostic feature of the region (Zelinsky 1977; Lewis 1972). Log houses and barns, once the single most common large farmstead structures in most areas, were selectively destroyed for many years because they were perceived to be obsolete and "*déclassé*" by many rural folk (fig 2.2). In more recent years outmoded outbuildings, corn cribs, outhouses, and smoke houses disappeared at an inordinately rapid pace, reflecting farmers' needs to demonstrate that they are practicing modern farming technology.

Fig. 2.1. Pennsylvania's rolling piedmont shown here in Lancaster County contains one important tangible identifier of the present-day Pennsylvanian homeland: The single-family farmstead dominated by a large barn with nearby two-story masonry house and relatively few outbuildings. In the nineteenth century, houses built of log and a greater number of outbuildings were more common. The Pennsylvania town, with its red brick buildings often built next to each other and flush with the sidewalk, characterize the region's urban counterpart. Photograph by Richard Pillsbury, 1984.

The Pennsylvanian barn is the center of the classic regional farmstead and is the focus of more research than all of the remaining buildings combined (Dornbusch and Heyl 1958; Glass 1971; Glass 1986; Ensminger 1992) (fig. 2.3). The most common version is a two-story structure with a stable and pad-dock attached to the first floor at the rear and hay mows and threshing floor on the second floor. The upper floor thrusts over the paddock by four to ten feet, while being accessed at the front by means of an earthen or stone ramp. None of these elements is standardized. Dornbusch and Heyl recognized 12 versions of this barn; Ensminger (1992) more recently identified 18 subtypes.

The Pennsylvanian barn dominates the farmstead by force of size, often being several times larger in cubic feet than the house and other structures.

Initially built of log or stone, surviving examples of these barns may date as early as the 1720s, though most are of nineteenth-century vintage. It is not clear whether adoption of this barn form in the late eighteenth and nineteenth centuries is owed to a growing Pennsylvania Dutch ethnic impetus or to a shift from subsistence to commercial grain and livestock farming, but the two-story barn became the regional standard by the beginning of the twentieth century. Two-story Dutch barns and single-floor English barns (with much the same functional arrangement as the upper floor of the two-story barn) dot the flatter areas of West Jersey and Delaware.

The Swensen brothers owned the first European house within the limits of Philadelphia, a Swedish log structure of unknown characteristics. Though not familiar with log construction technology at the time of their arrival, the

Fig. 2.2. Log buildings and outbuildings dominated almost all of the Pennsylvania homeland throughout most of the nineteenth century, but have almost disappeared in most areas today. Juniata County, Pennsylvania. Photograph by Richard Pillsbury, 1977.

Fig. 2.3. The classic Pennsylvania farmstead characteristically has a large farmhouse dominated by an even larger barn. Franklin County, Pennsylvania. Photograph by Richard Pillsbury, 1971.

English quickly adapted this technology to their needs. Scandinavian and Germanic room arrangements quickly lost favor to the single- and double-room arrangements common in England, Wales, and Ulster. The central chimney "continental cabin," as Kniffen and others have termed it, was rare—a museum piece today—so uncommon that Glass (1971, 181–83) was drawn to comment, after finding fewer than 40 in his 20-year study of the homeland, that "even the Pennsylvania Germans, for whom cultural purity is often alleged, have been synthesized through the processes of diffusion, adoption and change" (fig. 2.4).

Frame and masonry houses appeared from the beginning in small numbers, the first being more European than American. Several houseforms came to dominate the rural landscape, most notably a seven- or eight-room, two-story house, usually of masonry construction, with one or two front doors. Several proto-versions of this house of Germanic and British origin were introduced, but by 1820 they had evolved into a regional house with central hall, seven rooms, and two stories that had only regional dress to differentiate it from similar houses built elsewhere along the Eastern Seaboard at the same time. This two-story, two-room-deep Georgian home was both a common farmhouse and urban house throughout central and western Pennsylvania (fig. 2.5).

British settlers simultaneously introduced the English farmhouse known in America as the I-house after Kniffen's designation (Barley 1961; Kniffen 1965) (fig. 2.6). This house was common to the entire Eastern Seaboard, but in the Pennsylvanian homeland it almost immediately took on a regional dress

Fig. 2.4. The central chimney "continental" house of the Pennsylvania Dutch was comparatively uncommon on the landscape. Lebanon County, Pennsylvania. Photograph by Richard Pillsbury, 1977.

Fig. 2.5. The deep house with all of its characteristics is one of the most important diagnostic landscape elements of the Pennsylvanian homeland, though its complex history makes its identification with any single ethnic group difficult. Lancaster County, Pennsylvania. Photograph by Richard Pillsbury, 1988.

that included a preference for masonry construction, a somewhat squatter general appearance, and a second-floor, evenly spaced five-window fenestration. The I-house lost its ethnic identification as it was carried west, where it, with the double-pen log house and the two-story, two-room deep Georgian house described previously, became the third standard farmhouse of the region.

A variety of farmstead outbuildings, most notably the spring house and less commonly the smoke house, characterize the homeland. The spring house, a small shed built of log, stone, brick, or wood over a spring for keeping foods cool, usually had a stone foundation. Most smoke houses and outhouses have disappeared. Pig sties, chicken houses, granaries, equipment sheds, and the like rarely existed here as their functions typically were included within the overarching barn structure. The absence of ancillary structures is one of the defining characteristics of this farmstead.

The dominant multistory stone or frame barn, a large two-story masonry house, and a minimal number of outbuildings may indeed have been the most common Pennsylvanian homeland farmstead, but it was not the only farmstead assemblage—nor even the dominant farmstead morphology in all areas (Glass 1971). Log houses and barns existed throughout almost all of the region during most of the nineteenth century, including the relatively "English" Delaware and West Jersey areas. Kalm (Benson 1966) and other travelers noted hay ricks, barracks, and other outdoor hay storage technologies. A variety of smaller barns also seem to have been comparatively common, though few remain today. Farmsteads in central and southwestern Pennsylvania and

Fig. 2.6. The I-house originated in Britain, but was widely built by homeowners from a variety of backgrounds in central and western Pennsylvania and Maryland. Berks County, Pennsylvania. Photograph by Richard Pillsbury, 1974.

western Virginia built of log construction by first-generation settlers have all but disappeared.

The Pennsylvanian town took on a distinctive regional look almost from the beginning (Pillsbury 1968). Penn's use of a grid street pattern for Philadelphia set the tone. Lancaster, Carlisle, and other early seats of government also had this rectilinear street pattern. Distinctive in their street geometry, these places were notable as trading and artisan centers. The reason for the disproportionately large number of artisans and merchants in this region is not known, but the village of farmers with few commercial functions was never a part of this landscape. Commercial demand created market squares, and narrow building lots promoted greater access to commercial streets.

Thus, ethnic mixing and compartmentalization led to a patchwork quilt of milieus rather than a homogenous landscape of farmsteads, fields, and towns (Pillsbury 1977, 1987). Some early settlement areas had Germanic landscapes, others British landscapes, but in time ethnic landscapes merged into a regional standard exemplified in the later landscapes of central Pennsylvania west of the Susquehanna River and south of the Allegheny Front. The alternating ridges and valleys of central Pennsylvania allowed some compartmentalization, and it is possible to distinguish areas that had more Pennsylvania Dutch, English, or other settlers, but even their landscapes became increasingly homogenous in time.

The Pennsylvania Culture Area

Geographers and others have long recognized a "Pennsylvania culture area." Most maps of the area show its core to lie between the Delaware and Susquehanna in southeastern Pennsylvania in an area of intense early settlement by immigrants emanating from a host of European origins (Pillsbury 1968, 1987; Glass 1971). Philadelphia was the dominant center of innovation for this region, though Baltimore became an increasingly important entrepôt for new settlers as trans-Atlantic fares declined for passage on the largely empty returning tobacco ships (Glass 1971). The colonial foundations of the Pennsylvania culture area were shaped by the unique agglomeration of three dominant groups of emigrants who intermingled to create a distinctive way of life and sense-of-place identity. Though greater Philadelphia was the earliest settled part of the culture area and the hearth of a distinctive regional way of life, it is only nominally the core of the region today (Pillsbury 1987).

Central Pennsylvania, lying between the Allegheny Front and the Susquehanna, is often ignored or discounted by ethnic historians in their attempts to delineate a Pennsylvania culture. But this area in many ways is the truest expression of the region's culture. Assorted British and continental cultural elements fused here into an American—or more especially Pennsylvanian—reality in which ethnic origin plays an ever decreasing role in shaping the home one built, the food one consumed, and the beliefs one held. Certainly everyone who lived here knew his/her ethnic roots, but the accuracy of that presumption became less realistic as time passed, because of intermarriage and cultural assimilation.

The poorly defined southern border of this region is differentiated from the whole by the increasing role of Chesapeake Bay settlement influences in the landscape. The Shenandoah Valley, virtually all of western Maryland, and parts of Virginia adjacent to these areas have such strong identities that they are included in every attempt to define a Pennsylvanian culture, yet a careful analysis of their traditional visual landscapes and lifeways clearly differentiates them from southeastern Pennsylvania.

Structure of the Pennsylvanian Homeland

The structure and extent of the Pennsylvania culture area is not synonymous with the Pennsylvanian homeland. The Pennsylvania culture area is largely an *externally* defined device used to identify the distribution of a wide range of material culture artifacts covering a broad area within the Middle Atlantic states. In this sense the culture area is differentiated from the homeland in that it chronicles the presence of cultural associations in an almost timeless space. Relict and contemporary features are intermixed with scant concern for temporal evolution of a place and its cultural associations.

By contrast, the homeland is an *internally* defined area created by a sense-of-place association held by its inhabitants. Both internal and external forces may either shrink or expand the homeland's areal extent. The Pennsylvanian homeland, for example, has been shrinking for more than a century. The Industrial Revolution brought tens of thousands of emigrants from eastern and

southern Europe to Pennsylvania's western coalfields, shattering any homeland concept there. More recently the sprawling suburban field of Philadelphia has swept across southeastern Pennsylvania and portions of New Jersey
to obliterate any sense of traditional homeland association that may have existed in those areas.

The complex ethnic history and compartmentalized settlement of the
Pennsylvanian homeland created a distinctive way of life and sense of community, but was always characterized by a weak sense of self-identity as a holistic place. Its inhabitants, and more important their chroniclers, have clung to
the myths of the purity and uniqueness of ethnic heritages and identities that
have tended to obscure the region's overall commonality. Even today ethnic
writers examining the Pennsylvania Dutch cultural milieu hold tenaciously to
the role of ethnic purity. Ensminger's recent study of the Pennsylvania barn,
for example, identifies three classes, 13 types, and 18 distinct "Pennsylvania"
barns—all Germanic or Pennsylvania Dutch in origin (Ensminger 1992). Ensminger summarily dismisses the two-level English Lake District barn and an
assortment of single level non-Germanic barns found in large numbers in parts
of his region. This comment is not meant to disparage Ensminger's excellent
work, but to point out that cultural geographers and ethnic historians find it
difficult to separate the desire to trace ethnic heritages from the need to treat
a place as a coherent whole that transcends ethnicity. In a sense, the homeland
is better seen as an entity by outsiders than by its insiders.

There are some areas of the Pennsylvania culture area that clearly cannot
be considered to be a part of the homeland. The English core area of Philadelphia and its surrounding counties has been the most obliterated through time.
Like most entrepôt areas, this one was a significant transaction center for in-
coming ways of life and alternative lifestyles. Philadelphia, the region's capital, was a diverse city that reflected the whole region with about the same percentage of German population as the state as a whole (Purvis 1987). The
diversity of the city with merchants and artisans from many origins ensured
that new innovations and patterns of all sorts would be diffused to the population as a whole, not just those of like origin. The economic and political
power of the city, coupled with the actual diversity of the so-called English
counties, meant that its citizens had access to a wide spectrum of acculturated
innovations. Even a causal examination of Elizabeth Ellicott Lea's 1852 cook-

book, for example, clearly demonstrates that this Maryland Quaker housewife with few apparent associations with the Pennsylvania Dutch life, had access to an amazing number of German recipes; indeed her book well illustrates the acculturated Pennsylvanian diet—including a predictable English bias toward her own origins (Weaver 1982).

Moreover, Philadelphia's tentacles into its hinterlands ensured that it set the pattern for the region as a whole. Philadelphia's role as transaction center ensured that local decision makers and arbiters of taste had either direct or indirect contact with the city and much of the continuity of the region. The result was that the English counties of Pennsylvania and those adjacent parts of Maryland and New Jersey presented a somewhat different landscape look but overall were as much a part of the homeland as anywhere else in the eighteenth and early nineteenth centuries.

The economic growth that accompanied Philadelphia's role as transaction center also meant that ultimately the city and most of the surrounding area would just as certainly be drawn out of the homeland. The expansion of the region's industrial base after 1840, the urbanization of Philadelphia and many smaller communities, and the influx of a broad range of immigrants from throughout eastern and southern Europe separated the city from the homeland. The suburbanization that accompanied the rise of the truck and automobile after World War II inevitably removed more and more of southeastern Pennsylvania from the homeland.

Southern New Jersey's early Dutch, Swedish, and Quaker heritage meant that it was destined always to be a step child of the Pennsylvanian homeland with poor connectivity to the main region. The subarea's strong connectivity in architecture, speech patterns, and food preferences, but an almost complete lack of the Germanic elements of the classic Pennsylvania Dutch way of life, virtually condemned it to a peripheral role in any classification scheme. Clearly any attempt to depict a regional culture would have to include this area, at least as a subsidiary to the main body, but just as clearly it never interacted in the ways that would bring it into a homeland concept.

Western Pennsylvania and adjacent Maryland and Virginia were settled almost entirely by emigrants from the core homeland and under normal circumstances would be included as an integral part of the homeland, except that much of the region is underlain by some of the world's finest bituminous coal seams. If it were possible to resurrect an 1840 landscape in this now-devastated place there would have been such a strong resemblance to that found to the east that no lines of demarcation could be drawn, though Virginian influences could have been increasingly felt south of Washington and Uniontown, Pennsylvania. But this scenario was not to be and the invasion of hundreds of thousands of miners into the region changed that place for all time. Though mingling between the farming and industrial communities was minimal, and often antagonistic, the region can no longer be included in the homeland today.

The Past, the Present

The ethnic clutter of the Pennsylvanian homeland could all-too-easily lead one to define it solely in terms of ethnicity, but this simplistic solution overlooks the obvious intermixing that took place there in the eighteenth and nineteenth centuries. Attempts to define this area using ethnic presence have missed the point. What makes this place a homeland has little to do with ethnic origins. Rather, a lifestyle and an attachment to place that evolved over the years define the homeland and its residents. At the heart of these homelanders' values is an attitude that the land is to be farmed fastidiously and that all should respect their perceived ethnic origins. While the foregoing discussion has focused on the evolution and character of this distinctive regional landscape, the landscape is an end product of the presence of the homeland, not the homeland itself. As such it can never be more than a surrogate measure of the Pennsylvanian homeland and does not represent the homeland in its entire complexity.

Continuity has been an overriding theme here as few new people have entered over the last century or so and those who left did so only reluctantly. The Amish and Mennonites are a classic case in point. Stable in their location for several centuries, in the last few decades they have departed in significant numbers. Pressures upon them to leave—rapidly escalating land prices, constant intrusions by gentiles, gross commercialization, and broad exploitation of their lifestyles—would have propelled a less-determined people to new areas much more quickly. Yet in their colonies in Ontario, western Pennsylvania, northern Indiana, and elsewhere these Pennsylvania Dutch still have ties to the region and speak longingly of it. The tenacious few who remain may also succumb eventually to push factors.

The Future

The Pennsylvanian homeland is increasingly becoming a mere caricature of its former self. After decades of economic and population stagnation, unparalleled growth is now taking place. Lancaster County has become the most productive agricultural county in the northeastern United States. Manufacturing growth has spilled out of the Philadelphia area across southern Pennsylvania to the Susquehanna. Contemporary southeastern Pennsylvania, one of the largest centers of cabinet and specialized wood products manufacturing in the nation, supports a growing furniture industry and has become an important warehouse and transshipment area for the Eastern Seaboard. Suburbanization has exploded across the region with all of its attendant ills. New residents now flock to the area because of its strong economy, while urban foragers from throughout the Northeast have made it one of the most important rural tourist destinations in the nation. Traditional farming is rapidly being replaced by intense production of fresh produce and meat for an urban market. Roadside sales today have become so important in Lancaster and Chester counties that some local farmers are forced to import "farm fresh" produce and processed "home" goods from fringe areas to meet demands.

Few growth areas have been able to continue their homeland traditions in the face of mounting nationalization, and this region is no different. The nationalization of the region has been taking place at a rapid pace. A generation ago one could encounter at least one older resident with little or no English proficiency; today such encounters are rare. Even the Pennsylvania Dutch accent is disappearing among the youth, while regionalisms are becoming more a learned response than a natural outgrowth of heritage. Each day brings the region a bit more into the American mainstream.

CHAPTER THREE

Old Order Amish Homelands

. . . .

Ary J. Lamme III

To understand Amish geography requires recognition of one primary constituent of their lives—community. Submersion of the individual within the group is a key ingredient for a happy successful Amish person. Among themselves the Amish speak Pennsylvania Dutch and use various German words to convey ideas related to their common welfare, both spiritually and in a human sense. *Gemeinde* is the redemptive spiritual community within which Amish collectively seek to do the will of God (Hostetler 1981, 6). *Gelassenheit* is calm submission to God's will and God's way as interpreted by leaders (Kraybill 1989, 25). The Amish are gathered together into a community of believers, acting in concert here on Earth while awaiting ultimate redemption in the hereafter.

This concept of the community is of utmost importance, for it has enormous spatial implications. Pulling together into a close-knit group, separating from worldliness, while establishing everyday life practices that purposely exacerbate differences between themselves and the rest of the world, creates distinctive patterns susceptible to geographic analysis. Those buggies filled with distinctly dressed individuals rolling along country lanes are symbolic of a lifestyle closely tied to the land.

Old Order Amish Origins

Old Order Amish religious practice evolved from the Anabaptist movement of the early Protestant Reformation (Hostetler 1993). In 1525, less than ten years after Martin Luther first challenged Roman Catholic practice, a small group of Protestants began adult baptisms. Religious and civil authorities immediately persecuted this first group of Anabaptists, or rebaptisers, because they had been originally baptized as infants in the Catholic Church. Alienation from society and willing martyrdom in service to their cause are concepts that originated in those times and remain a distinguishing feature of Old Order Amish faith.

In the 1530s an influential Anabaptist leader emerged in Menno Simons, a former Catholic priest. Eventually many Anabaptists came to be called Mennonites because of the writing and influence of Simons. Anabaptist beliefs from that time have been summarized as: literal obedience to Christ, the church as a community of believers, adult baptism, separation from the world, exclusion of those who fail to uphold the faith, rejection of violence, and refusal to swear oaths (Kraybill 1989, 5). These fundamental tenets, interpreted in a contemporary context, guide Old Order Amish life today.

Mennonites, as the followers of Menno Simons called themselves, spread through many sections of Protestant Europe during the next two hundred years. Present-day Old Order Amish in North America draw their name from a contentious Mennonite leader of the 1690s named Jacob Ammann. Although little is known of him personally, and the Amish have little interest in stressing the role of individual leaders, Ammann insisted on severely conservative interpretations of Anabaptist doctrine, in particular, harsh punishment of the wayward, conservative appearance, and extremely strict discipline within the group. His ideas led to a schism with other Mennonites, creating a group originally called the Amish Mennonites. The Amish applied the designation "old order" in North America to indicate differences with several less conservative Amish groups.

Mennonites began coming to North America around the time of the Amish schism. Many settled in southeastern Pennsylvania and became the foundation of so-called Pennsylvania Dutch society. The first Amish Mennonites are thought to have arrived in 1727 and settled in Pennsylvania as well. Over the next century approximately two thousand Amish arrived from Europe to form the basis of Old Order Amish society in North America. Today the Amish are simply the most conservative theological branch of a much larger Anabaptist Mennonite religious group in North America. It is the conservative characteristic of these particular Anabaptists that leads them to insist on doing things in ways radically different from the rest of society, such as driving horse-drawn buggies without slow moving vehicle emblems.

Amish Homelands

In absolute numbers, the Old Order Amish are not an imposing group. Even in the townships where they have significant numbers, they remain a minority population group. In Lancaster County, Pennsylvania, perhaps the area most widely identified as Amish, they number only approximately 15,000 out of a total population of around 450,000 (Kraybill 1989, 9). Even among Mennonite affiliated religious bodies, the Amish account for only one out of seven adherents. The Amish and other members of "plain" churches tend to stand out in Lancaster County because of their distinctive garb and way of life. In strictly proportional terms then, the Amish may never qualify as possessors of a homeland.

When it comes to place, Amish areas tend to be located in mid-latitude, rural agricultural regions (map 3.1). Longer-settled areas include parts of the most prosperous sections of the Corn Belt and the Middle Atlantic states. Newly settled areas tend to be marginal agricultural regions where the Amish can acquire relatively inexpensive small farms in close proximity to one another (fig. 3.1).

Using distinctive agricultural practices, the Amish purposefully create a distinctive cultural impress on the landscape. For instance, the Amish stack corn shocks in their fields by hand, much in the manner common in rural areas of the United States before World War II. The absence of modern mechanized equipment also gives the Amish farm a distinctive look. Most important, the Amish sense of place (as a minority people living within the larger society) requires a sense of sharing an area with others. The Amish cling to their way of doing things as most right and recognize differences between themselves and others but with relatively little condemnation of nonbelievers. They live a life of peaceful coexistence and cooperation.

Old Order Amish farms have existed in the United States for 250 years.

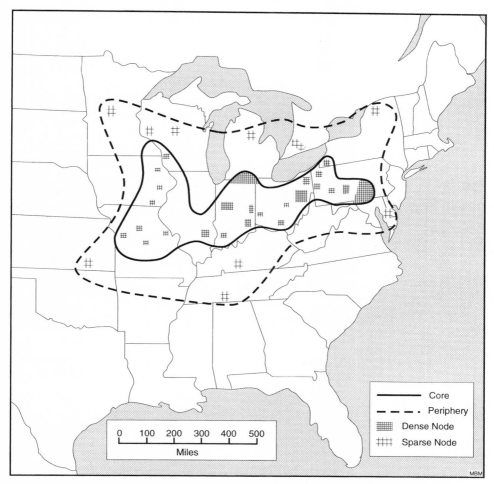

Map 3.1. Old Order Amish homelands (2000) are located primarily in the northeastern United States. This map distinguishes between the Amish core, where church districts are numerous in established nodes, and the periphery, where nodes containing church districts are recent and relatively few. High birth rates and high church membership retention have prompted an outward diffusion. Single church districts located in eastern states and Montana are omitted. *Source:* author.

Fig. 3.1. A typical Old Order Amish farmstead in northern New York. This was a marginal farm with old outbuildings and silos prior to its purchase by the Amish. The farmhouse, lacking modern conveniences, is perfect for a large Amish extended family. Photograph by Ary J. Lamme III, July 1996.

They are recognized as important elements of the cultural landscape in those long-settled areas, and in some cases, they are the basis of a tourist industry. On the other hand, Amish farms in newly settled areas such as the North Country of New York are widely known to attract visitors as well. Thus, length of residence does not appear to be a significant element distinguishing Amish settlement areas.

When considering the element of control, the Amish are certainly unusual. In fact, their relationship with civil authorities is one of overt submission while maintaining their beliefs. The Amish avoid normal expressions of power. They do not seek elective office, they decline government and military service, they refuse most government assistance, and they avoid taking any sort of evangelistic stance about their beliefs or lifestyle. They will participate in the political process in a low-key way if issues that concern them are at stake.

Land-use regulation or planning of new highways or facilities in their area prompt the Amish to seek special ways to express themselves. In terms of temporal control, however, land ownership of many farms within a restricted area is the extent of their range of power.

It is with "bonding" that the Amish have the greatest claim to a homeland. They have a very high degree of attachment to the land based on biblical support of agricultural activities as well as a desire to separate themselves from the densely settled areas of nonbelievers. Separation from others has both physical and mental expressions. In either case, it requires an especially strong sense of identification with a particular place. These are the Amish homelands. However, their love of the land is not unlike that of many farmers and others who cherish the expression of pastoral values. The extent of Amish bonding with their homeland, and its spatial expression in their thinking, are revealed by a formerly Amish man who said, "To an Amish person the 'world' begins at the last Amish farmhouse" (Wittmer 1990, 9).

Amish Religious Communities

The geography of the Old Order Amish in North America is based on religious communities. In a number of cases the Amish have failed in their attempts to create such communities (Crowley 1978). In others, they created permanent settlements. These successes and failures, indeed of the entire geography of the Amish in North America, is linked to the Amish sense of community, and an essential ingredient of these communities is their affective spatial expression that we can accurately term homelands.

Hostetler devotes a chapter of the fourth edition of his classic *Amish Society* to a discussion of the community (1993, 91–113). Although the Amish sense of community has its material aspects, he points out that it is basically a mental construct: "As a corporate group in the United States and Canada, the Old Order Amish celebrate communion and break bread together; they represent a community of one mind, one discipline, and one body" (91). Part of this mental sense of community is bonding with land, an important element in the definition of homeland.

Hostetler says that the Amish community consists of three elements found throughout North America. Each of these elements has a spatial expression. The first is the settlement area that includes all Amish families living in relative proximity to one another. They will normally be interspersed among non-Amish families (all who are not Amish are called "English"), but within reasonable buggy-riding distance from one another.

The second is that the Amish gather on different individual farmsteads every other Sunday to hold religious services. Because homes and other farm structures can hold only a limited number of people, the Amish divide their settlement areas into church districts consisting of 25 to 35 families, a grouping that becomes the basic unit of Old Order Amish society. In the early 1990s there were approximately eight hundred church districts in North America containing a total Amish population of about 130,000 (Hostetler 1993, 97). Ohio has the most districts (over 250) while a number of states have only one.

The third aspect of community is affiliation, a group of church districts that hold to the same theological beliefs and practices. Although the home church district regulates the life of the individual, Amish can attend services and maintain fellowship with other districts of the same affiliation.

Non-Amish tend to think of the Amish as a homogeneous group. Non-Amish notice the plain garb without distinguishing between different forms of the dress. Of course, the Amish themselves try to submerge individual differences within the religious community. To understand the Amish, however, is to realize that the superficial homogeneity we observe masks important differences within the Old Order Amish community and between them and their Mennonite brethren.

Founded through doctrinal schism, the Old Order Amish have experienced such splits themselves. Over the past hundred years a number of groups have broken away from the Old Order Amish to form other Amish groups. Within the Old Order there are differences of interpretation as well. In this patriarchal society, leaders of individual church districts decide how life is to be lived for those within the community. Schisms within the Old Order Amish have almost always been over questions relating to accommodation with the modern world. Because the Amish must maintain their separate communities, yet live within a larger nonbelieving society, such decisions are vital to maintaining group identity.

The Old Order Amish in their varied forms are diffusing in North America. Wherever they go they carry with them this sense of community consisting of attraction to areas where other Amish live, membership in the church district, affiliation with other church districts, separation from the "English" world while struggling to be accommodated within it, love of the land and agricultural activity, and spatial circulation within the settlement area. In fact, when the Amish want to move to establish a home they have a pretty good sense of the range of their spatial options.

Amish Core-Periphery Distinctions

Most graphic representations of the Old Order Amish in North America have been dot maps (Lamme and McDonald 1984). Using dots of varying sizes to represent the number of church districts in different settlement areas has provided a reasonably clear picture of Amish locations. But the concept of homelands combined with the distribution of church districts gives a more insightful Amish areal representation (map 3.1). Nodes in this map show areas where multiple Amish church districts are located. Dense nodes in the core contain more church districts, while sparse nodes in the periphery contain fewer church districts. Various pressures are forcing the Amish to relocate from the more desirable core to the less desirable periphery.

The Amish homeland core centers on the climatically moderate agricultural regions of the Middle Atlantic states and the Midwest. Here the Amish have had settlements since the first half of the nineteenth century. Population pressure in the Amish core, a result of high birth rates and high membership retention, has encouraged Amish diffusion to the periphery. When possible, the Amish choose to relocate within the core. But a need for numerous contiguous farms that are the target of Amish group migration, and somewhat limited amounts of capital to invest in new land, have forced the Amish to look outside the core. The periphery, with its hillier, less fertile soil and harsher climates toward the north, is an obvious possibility. The North Country Amish settlements of northern New York are an example of this dispersal to lesser endowed, cheaper farmlands (Lamme and McDonald 1984, 1993).

Settlements in the periphery then are of more recent origin. A climate of Amish acceptance in the homeland core has spread into the adjacent periphery. The Amish have an overwhelmingly positive public image in the northeast, yet their diffusion is not accomplished without challenges. Conservative rural areas experiencing the rapid influx of a large group of unusual people who resist assimilation, refuse to send their children to school, and drive horse-drawn vehicles that leave behind manure are unlikely to win immediate popularity among local inhabitants. Yet, through a combination of humble stubbornness, compromise, and general good will, the Amish have managed to be successful in new areas in their spread to a homeland periphery.

Conclusion

The Old Order Amish are unlikely to meet the demographic, temporal, and political-control criteria that establish people in other areas as possessing homelands. Their numbers are too few and their distribution is too dispersed for the Amish to have a comparatively significant presence. On the other hand, in the category of intensity of feeling for place, the Amish qualify hands down as possessing legitimate homelands. Standards that qualify an area as a homeland should weigh heavily the intensity of the feeling factor.

CHAPTER FOUR

Blacks in the Plantation South

Unique Homelands

. . . .

Charles S. Aiken

The future of American Negroes is in the South. Here they have made
their contribution to American culture. . . . Here is the magnificent
climate; here is the fruitful earth under the Southern sun; and here, if
anywhere on earth is the need of the thinker, the worker, and the dreamer.
—W. E. B. Du Bois, 1947

During the civil rights era (1954–72), Jerry Lewis, a well-known comedian,
appeared on the *Jack Paar Show*. Fumbling for humor, Lewis remarked that,
every time he was on an airplane that flew over Mississippi, he made a special
effort to flush the toilet. Lewis's intent was to suggest that white Mississippi-
ans, as racial bigots, should be defecated on. Letters of protest poured into
NBC's New York headquarters and into affiliate stations from people who
charged that the remark was slanderous to their state. To the chagrin of Lewis
and NBC, much of the outcry over the remark was not from whites but from
blacks (Johnson 1996). Lewis was hardly the first or the last to assume what
James Cobb observed to be the habit of "identifying Southern whites as
'Southerners' and Southern blacks as 'blacks'" (Cobb 1996, 10).

Some scholars of the South have unintentionally fallen into the same trap.
Even the distinguished historian C. Vann Woodward appeared at times to for-

get blacks. To Martin Luther King Jr. and other civil rights workers, Woodward's *Strange Career of Jim Crow* (1955) was a bible of the civil rights movement. But in another of Woodward's books, *The Burden of Southern History* (1960), blacks can hardly identify with two of Woodward's three ways in which the South's history differs from that of the nation. Members of old southern white families might relate to his argument that the United States history of success, innocence, and economic abundance is not the history of the South, but southern black families view the interpretations from a different perspective. For blacks, the defeat of the Confederacy is a story of victory, not one of loss. Blacks were the victims of slavery and segregation, hardly the ones condemned to contend with its guilt. Only in poverty did the history of southern blacks and whites converge, but even there, what economic abundance was found in the South belonged far more to whites than to blacks.

Much of the current argument over continuity in southern history and the South's loss of distinctiveness and sectional identity is conducted from perspectives that are largely irrelevant to blacks. If blacks are remembered, even recognized as a central component of the South, it is often within the context of essentially passive roles. The omission of blacks is found especially in the prattle from the neo-Confederate right, which includes people who among other things fought to keep the Confederate States of America battle flag flying over the capitol buildings of Alabama and South Carolina and defend its attachment to the state flags of Georgia and Mississippi ("Civil Rights Memorial Shadowed by Rebel Flag" 1989; McAlister 1990; Auchmutey 1993). The spurning of blacks in definitions and discussions of the South and what is southern lends credence to repeated complaints by African Americans that they and their rich culture are largely oblivious to whites (Du Bois 1903, 181–82).

The Plantation Regions as Homelands

For most of their history in North America, blacks were a rural people largely confined to the southern plantation regions (map 4.1). At the beginning of the twentieth century, 90 percent of the nation's blacks lived in the South, and 84 percent of them lived in rural areas, concentrated in the plantation regions. These regions, it should be emphasized, were not "homelands" to which blacks freely emigrated. As captives, blacks were torn from their African homes and

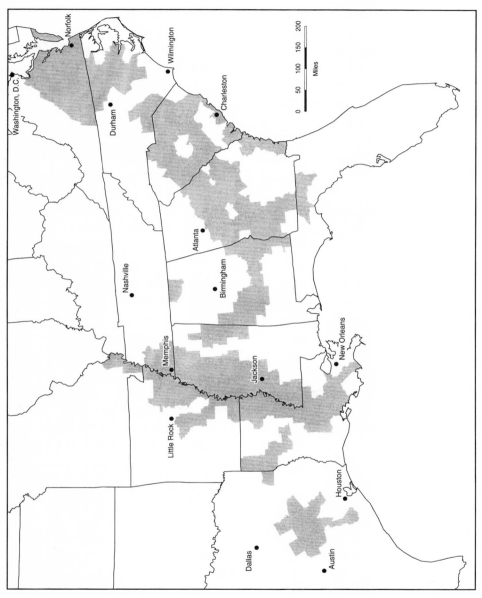

Map 4.1. Major historic and contemporary plantation regions, 2000. The black population ranged from 25 to more than 75 percent across these areas. *Source:* Based on Aiken 1998, 6.

forcibly sold into the plantation areas as slaves. Even after the Civil War and emancipation, the migration of blacks from older plantation regions such as the Lower Piedmont to expanding plantation areas such as the alluvial Mississippi Valley was largely manipulated by white planters who needed inexpensive black labor.

White industrialists also played a critical role in the migration from southern plantations to the manufacturing cities of the North. Only after the First World War and restrictive laws in the 1920s curtailed cheap foreign immigrant labor did jobs begin to develop. Northern industrialists went so far as to recruit blacks through newspaper advertisements and labor agents (Johnson and Campbell 1981, 79–88).

By 1970 the black population in the South had dropped to 53 percent and only 18 percent of the nation's blacks were in the rural South. Most of the redistribution of blacks occurred during the brief 40-year period between 1920 and 1960. In 1990, 52.8 percent of the nation's blacks lived in the South, including 3,474,582 who were rural (U.S. Bureau of the Census 1992). The decrease in actual numbers of blacks in the nonmetropolitan South was not as great as the relative decline. At the end of the twentieth century, several million blacks still resided in the old southern plantation regions and were the dominant or a sizable majority population in the counties that comprise the old plantation regions.

The emigration of blacks from the plantation South is presented by some scholars as having been initiated by blacks, a carefully planned escape from racial oppression and poverty (for example, Mandle 1978, 82–83, 84–97; Cohen 1991, xi–xvi). That the migration was actually a traumatic severing of a people from a place by economic, social, and political forces is rarely considered. Many blacks, even when racial oppression was at its worst, had no desire to leave the rural South but were forced to do so in the wake of collapsing plantation economies and the mechanization of plantation agriculture (Aiken 1998, 63–132). When mechanization threatened eviction from Rainbow Plantation in Tallahatchie County, Mississippi, in 1955, Nathan Kern, a tenant farmer, told Billy Pearson, the owner, that "Those of us who are still here, we're here because we chose to be here. We don't have to stay. We've all got cousins and kin up north and all we have to do is send a postcard saying save me a room, and we're gone. We stayed because this was our home and now we wonder if it's our home any more" (quoted in Halberstam 1993, 453).

Richard Wright, whose family migrated to Memphis and then to Chicago from a plantation in Adams County, Mississippi, because of the decimation of cotton by the boll weevil early in the twentieth century, personally experienced the great loss blacks encountered in leaving the rural South: "We look up at the high southern sky and remember all the sunshine and the rain and we feel a sense of loss, but we are leaving. . . . We scan the kind black faces we have looked upon since we first saw the light of day, and, though pain is in our hearts, we are leaving" (Wright and Rosskam 1941, 92).

Not all blacks have nostalgic ties to the plantation South. Many metropolitan blacks, even some who live in southern metropolises, have no desire to visit, much less return, to the plantation regions. Anthony Walton relates the story of his father, who left a plantation in Marshall County, Mississippi, in 1950 for Memphis, like Richard Wright a generation earlier, and then moved northward to Chicago. When Walton asked his father, "Tell me about what it was like when you were growing up," the reply was "Ain't nothing to tell." And when he asked, "What's the first thing you remember," his father's terse answer was "Bigotry." "I call it 'sippi' cause I don't miss it" (Walton 1996, 213–19).

Africa as "Homeland"

Because first slavery, then emancipation, segregation, and discrimination are associated with the South's plantation regions, as a group, American blacks look past the plantation South to Africa as their homeland. This view has led in recent years to growth in tourist visits to Africa by American blacks, who increasingly have become more affluent, and also to increased political pressure from them for the United States to expand its role in economic development and trade with sub-Saharan Africa (Brooke 1998; Phillips 1998).

Since before the Civil War, "back to Africa" movements have been supported both among blacks and certain groups of American whites. Liberia was created in 1822 by a United States colonization society as a country to which to send freed slaves (Redkey 1969). Marcus Garvey in the early 1920s led more than two million followers in the largest effort of the twentieth century to entice blacks to move to Africa. The project collapsed in 1925 after the charismatic Garvey was convicted of mail fraud. In the latter half of the century, especially during the civil rights movement between 1954 and 1972, various small groups and a few prominent individuals in disgust fled the United States for Africa. In the face of what he wrote in 1946 about the future of American Negroes in the South, W. E. B. Du Bois was among those who fled to Africa. Ironically, Du Bois died on the eve of the greatest mass protest by blacks, the March on Washington in August 1963 (Reed 1997).

Most American blacks who visit Africa or renounce the United States and move there usually do not find what they seek. Africa is a continent of Third World countries without the infrastructure or the unity to fulfill utopian visions. Health care is inadequate, mortality rates are high, educational levels are low, and famine and civil war continually plague the continent. Most of all, many African nations suppress personal freedom to a degree unknown in the United States.

A compromise to the "back to Africa" crusades, which grew out of the black power movement initiated in 1966, is to express African heritage through revival of particular aspects of African culture and expansion of African studies in American universities. Ironically, the emphasis on African heritage eventually led to proposals to replace the term *black*, which was originally part of "black power," with *Afro-American* and *African American*.

The Plantation Regions in Reality

Whether they call them "areas of origin," "homes," or United States "homelands," most African Americans are tied in some way to the plantation regions of the South. Few Americans have significant remembrance or connections to their heritage beyond four or five generations. This is true for whites and blacks. Even the late-twentieth-century fad of tracing one's ancestry, in part triggered by Alex Haley's book *Roots*, for most people ends at about one hundred and fifty to two hundred years ago.

Most African Americans who seek to trace their lineage eventually are forced to the plantation regions and to records that, surprisingly, are often more complete than those of the descendants of poor whites. Planters kept detailed accounts, even knowing the areas of Africa from which their slaves came and the groups to which they belonged. Many planters' records survive in families and legal archives (Ball 1998, 18–21; Gomez 1998). Blacks who attempt to ignore or who try to forget the plantation regions from which their parents, grandparents, or great grandparents migrated, for genealogical purposes even-tually are faced with them.

During the confrontational phase of the civil rights movement, many young urban blacks ventured into the rural plantation regions, which they re-garded as places filled with the shame of slavery. Their mission was one of cleansing the areas by abolishing segregation and restoring voting rights to the black inhabitants (Aiken 1998, 198–99). As Anthony Walton grew to man-hood in Chicago, he came to realize that though he had never lived in Missis-sippi and only rarely visited relatives in the state, he was, nevertheless, partly a product of the place: "My father's life was the Rubicon of my own imagina-tion. Yet for years I had hardly considered him a part of my story; rural Mis-sissippi had no clear ties to suburban Chicago. But I realized now that the one would not exist for me without the other" (Walton 1996, 212).

Commitment to the Plantation Regions

Because Christianity and the Christian church became so important to most blacks, many related their experiences in the plantation regions to various sto-ries in the Bible. Like Joseph in the Old Testament who was sold into slavery in Egypt, blacks were sold into slavery in the South. And like the Children of Israel, they awaited deliverance from bondage. The former slaves found, how-ever, that with emancipation they were still in the plantation regions, an alle-gorical "Egypt Land," and were still tied to the plantation by the new forms of servitude, farm tenancy and segregation. As late as the 1940s, black share-cropper families trudged to plantation fields at the pace of the African Amer-ican spirituals "Go Down, Moses" and "We Shall Be Free." The 1965 and 1968 Civil Rights Acts, together with the 1965 Voting Rights Act, symboli-cally delivered blacks from the final bondage in Egypt Land, giving them, among other things, freedom of movement and access to various kinds of jobs (Aiken 1998, 165–282).

To the majority of African Americans who now live in metropolises, the southern plantation regions, as for Walton, are not *actual* homelands but *re-membered* homelands or *perceived* homelands. With each urban-born genera-tion the memory of homeland recedes, and the rural South as a perceived homeland, which may be symbolic or illusionary, becomes more overt. But what about the several million blacks who still reside in towns, cities, and the rural countryside of the old plantation regions? Are these people really attached to the places? Do they consider the plantation South home, or do they still consider themselves sojourners in a foreign land?

Evidence of attachment to place is not always readily found. Stories and statements must be accepted cautiously, especially if they stem from some nostalgic utterance. When Morgan Freeman's professional acting career flourished, he returned to Tallahatchie County, Mississippi, to establish his home. James Morgan, a white who also had fled Mississippi, interviewed Freeman. To Morgan's consternation, Freeman spent most of the interview praising the virtues of Mississippi (Cobb 1996, 18).

Charlayne Hunter-Gault's return to the University of Georgia in 1988 to give the commencement address was hardly similar to her first appearance on the campus. As one of the first two black students to enroll at the university, on a cold night in January 1961, Charlayne Hunter was met by a white mob that shouted racial slurs and hurled rocks at the side and through the windows of Center Myers dormitory where she was housed. An Atlantan who long ago left Georgia for a distinguished career as a television journalist, Hunter-Gault responded to her warm reception at the university in 1988 with "It's good to be back home again. In a place that I have always thought of as 'our place'" (Hunter-Gault 1992, 248).

Although Hunter-Gault's statement about home was sincere, its full meaning was obscure to all but those who had lived through the tumultuous years of the civil rights movement and had witnessed the desegregation of the University of Georgia. The statement was also made by one who sincerely identified herself as a southerner but who long ago had ceased to be a real resident of the South. Hunter-Gault's southern home was a remembered rather than an actual one.

"Return" Migration

A more tangible measure of their attachment to the South is blacks' recent move to the South from the North and West. In the 1970s, the Bureau of the Census issued population estimates that suggested that the great emigration of blacks from the South was over, for more were moving to it than were leaving. The 1980 and 1990 censuses confirmed the trend. Unfortunately, the movement of blacks to the South has been interpreted by some scholars in the context of "return migration," implying that blacks who emigrated are now returning to the rural plantation areas which they left. One study even construed the return as a rightful "reclaiming" of the rural South by African Americans (Stack 1996). Most of the blacks who move to the South from the North and West were not born there, and their destinations are Atlanta, Birmingham, Memphis, and Charlotte, not small towns and rural areas. Kinship ties may be a factor in the migration, but the primary reasons for the moves are economic. Because only 7 percent of the blacks who moved to the South from other sections of the nation between 1990 and 1996 were age 65 and older, retirement was not a primary motive.

Among the blacks who migrated to the South between 1990 and 1996, 86 percent moved to metropolitan counties, 59 percent to suburbs. The Atlanta, Washington, Houston, Miami–Fort Lauderdale, and Dallas–Fort Worth metropolitan areas reported the largest gains in black population. Although arguments can be made that certain metropolitan areas of the South, including Atlanta and Memphis, have a plantation tradition, such a case can hardly be made for some southern metropolises, including Miami–Fort Lauderdale and Houston.

Farm Ownership

When former slaves began their journeys to freedom after the Civil War they were destitute. Few owned significant material possessions, much less farms or even a few acres on which to build a shack and scratch out a subsistence agricultural existence. Between 1865 and 1885 new systems of farming in the plantation regions employed most blacks in various forms of tenancy, ranging from closely managed sharecroppers to unsupervised cash tenants (Aiken 1998, 29–62). By 1900, the attainment of farm land ownership had become the most important measure of socioeconomic advancement of blacks in the plantation South, and, curiously, it remains a primary gauge.

The selection of farm ownership to measure economic advancement eventually proved an unfortunate choice, for it increasingly hid and took away from the diverse nature of a growing nonfarm black middle class in the plantation South. But the idea that blacks are essentially a peasant people who innately desire simple farm life but were driven from the land by discrimination and agricultural mechanization is still popular. Farm ownership and the number of black farmers are still used by some researchers as measures of socioeconomic progress by blacks in the plantation South (Aiken 1998, 345–49). A 1982 study by the United States Commission on Civil Rights reported that the rate of decline of farms operated by blacks was 2.5 times greater than that for farms of whites and concluded that "the urgency of this situation is accentuated by the virtual irreversibility of black land loss. . . . Today, only those who inherit land or who have other nonfarm sources of income can afford to purchase and operate farms. . . . The need for [federal] intervention is immediate" (quoted in Sinclair 1986).

Studies of the decline in the number of black farmers have two major problems. First, most fail to emphasize that the majority of black farmers were tenants who owned no land. Second, the decline in the number of black farmers in census statistics is not a measure of loss of land owned by blacks. Because a farm operation or "a place," the term on the schedule for the 1997 Census of Agriculture, is not the same as a cadastre, a landholding, census figures reveal the decline of land farmed by blacks but not loss of land owned by them.

In the first decade of the twentieth century, 75 percent of the black farmers in the South were tenants while only 25 percent were landowners. Tenant farming was not favored by either planters or former slaves. It developed as an expedient compromise which initially gave the former slaves the illusion that assigned tracts within a plantation were their own "farms." Because tenants paid for use of the land with a portion of their crops, the advantage to planters was that in a cash-poor South, farm labor did not have to be paid on a daily or weekly basis (Aiken 1998, 16–22).

Increasingly, poor whites as well as blacks were drawn into tenancy, and in 1930 tenants comprised 42 percent of the nation's farmers. Early in the twentieth century, a small group of scholars began to expose the social and economic evils of tenancy. Although tenancy was found in all sections of the nation, it was largely thought of and written about as a southern phenomenon. In 1930, 56 percent of the farmers in the South were tenants, but tenants also comprised one-third of the farmers in the Middle West (U.S. Bureau of the Census 1947).

Three geographical characteristics distinguished farms owned by blacks. They occurred in clusters of two or more, they were relatively small, and they usually had undesirable environmental traits, including inferior soils, poor drainage, or severe erosion. Blacks acquired farms in several ways. Some were sold to them by their former masters following the Civil War. Planters who fathered illegitimate children sometimes willed land to their offspring (Aiken 1998, 157–58).

Private and federal settlement projects created a few areas of farms owned by blacks. One of the largest private settlement projects was created by the Louisville, New Orleans, and Texas Railroad in the heart of the Yazoo Delta, a poorly drained, malaria-infested alluvial basin in northwestern Mississippi. Few whites wanted to buy land and live in such a place. In desperation railroad officials turned to blacks, who were perceived to relish heat and humidity and to be immune to malaria. Between the mid-1880s and 1904 the settlement grew to approximately twenty-five hundred blacks on numerous small farms focused on Mound Bayou, an all-black town of four hundred (Crockett 1979, 12–15). The resettlement and tenant-purchase farm programs of Franklin Roosevelt's New Deal also created farms for blacks as well as for whites. Gee's Bend in Wilcox County, Alabama, and Tallahatchie and Mileston Farms in Tallahatchie and Holmes Counties, Mississippi, are examples.

The plantation system that employed large numbers of black and white tenants began to disintegrate in particular cotton plantation regions with the arrival of the boll weevil. The insect crossed from Mexico into the United States in the 1890s, reached the Mississippi River about 1910, and by the early 1920s had entered southern Virginia, having moved across the expanse of the Cotton Belt. In the areas where plantation management was weak or essentially nonexistent, a few disastrous years of low cotton yields worked havoc with the agricultural infrastructure, especially with the financial sector (Aiken 1998, 63–96). The collapsing plantation system pushed blacks toward the viable agricultural regions of the South and toward cities. The mechanization of plantation farming, which began in the 1930s, together with federal crop acreage allotment and price support programs, destroyed traditional tenant farming in the viable plantation regions. A small number of black and white wage-workers, who drove tractors, mechanical cotton harvesters, and other machines, replaced the legions of tenant families (Aiken 1998, 97–132).

The tenant plantation system and low ownership of land among southern black farm families had a significant impact on the demographic changes in the southern population. For blacks who remained in the plantation regions, as for whites, a great shift from farm to rural nonfarm status occurred. In 1990, blacks operated only 43,487 (2.3%) of America's 1,925,300 farms, and only 117,083 (0.4%) of the nation's 29,930,524 blacks resided on farms, primarily in the South. Most of the South's black rural population is rural nonfarm.

Home Ownership

The residual impact of the plantation system is revealed in the ownership of black housing in 1960. Because most of the black farm households in the plantation regions were tenant families and much of the housing for blacks in plantation towns and cities was rental, home ownership among blacks was low. Low rates of home ownership persisted as blacks were severed from plantation agriculture. In 1960 only 22.7 percent of the nonwhite households in the Yazoo Delta plantation region of Mississippi and 25.4 percent in the Alabama Black Belt owned their homes, compared with the national average of 38.4 percent for nonwhites (U.S. Bureau of the Census 1963, I, pt. 5, 38–44; I, pt. 2, 65–69; I, pt. 1, 211). Similar levels of home ownership among nonwhites prevailed across the plantation regions.

The exodus of black tenant farmers from the land resulted in numerous vacant shacks scattered across the plantation landscape (fig. 4.1). In the Yazoo Delta, 12,971 dwellings, 12.5 percent of the region's housing units, were vacant in 1960, but only 1,933 (15 percent) were available to renters or buyers. In Tunica County, just 115 of the 1,107 empty units were for sale or rent (U.S. Bureau of the Census 1963, I. pt. 5, 38–44). The numerous empty houses awaiting demolition did not portend what was to happen.

Home ownership among blacks in the plantation regions today is strikingly higher than in 1960 (map 4.2). The War on Poverty in the 1960s made substantial funds available through the Farmers Home Administration and the Department of Housing and Urban Development to construct new dwellings for ownership and for rent. This is part of the explanation why, in the Yazoo Delta in 1990, blacks owned 47.6 percent of their dwellings; in the Alabama Black Belt the ownership rate was 65.1 percent. Across the old plantation regions the rate of home ownership is at least 43 percent, which is the national average for black-occupied housing units. Especially impressive are large areas of the plantation South in which the percentage of housing units owned by blacks is greater than the national average for all households. In the old

Fig. 4.1. The shotgun house is strongly associated with southern blacks. This boarded-up plantation example in Tunica County, Mississippi, symbolizes what happened to many black families in the twentieth century: By 1960 mechanized agriculture and agricultural decline forced most black tenant farmers from their plantation homes. They took jobs in urban areas in and out of the South. Today a significant number of blacks are moving into new nonfarm housing developments in and near municipalities in the plantation "homelands." Photograph by Charles S. Aiken, March 1961.

plantation regions of Virginia, the South Carolina coastal plain, the lower Piedmont, and the Alabama-Mississippi Black Belt, blacks owned 64 percent or more of their dwellings.

The large increase in home ownership indicates two important characteristics of blacks who remain in the plantation areas. First, it reveals that blacks have made significant economic and social advances during the latter part of the twentieth century. This advancement is directly related to the severance of blacks from the plantation system and to the civil rights movement that helped give blacks the rights guaranteed all Americans under the Consti-

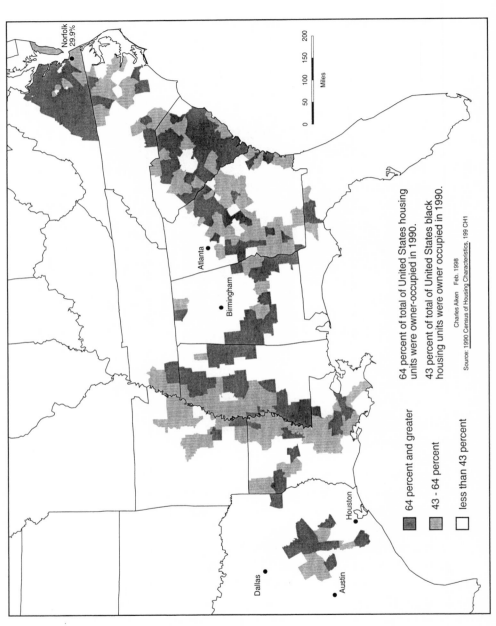

Map 4.2. Percentage of blacks who owned the dwelling in which they resided in 1990. *Source:* U.S. Bureau of the Census.

64 percent of total of United States housing
units were owner-occupied in 1990.

43 percent of total of United States black
housing units were owner occupied in 1990.

Charles Aiken   Feb. 1998

Source: 1990 Census of Housing Characteristics, 199 CH1

64 percent and greater

43 - 64 percent

less than 43 percent

Norfolk
29.9%

Atlanta

Birmingham

Houston

Dallas

Austin

0   50   100   150   200

Miles

tution. In 1960 farming and personal services led as black occupations in the plantation regions. By 1990 manufacturing and professional services had become the two primary occupations (Aiken 1998, 356–61). Second, the substantial increase in home ownership among blacks since 1960 is tangible evidence of the commitment of blacks to the plantation regions as home. For blacks, as for most Americans, investment in housing is usually the largest single financial obligation in a lifetime.

The increase in home ownership among black households has been accompanied by momentous changes in local population distribution and settlement patterns in the plantation regions (Aiken 1985, 1987, 1990, 1998, 307–39). Blacks now live in hamlets in the countryside, within black residential areas of municipalities, and on the margins of particular municipalities.

Tallahatchie County, Mississippi, illustrates the new settlement patterns in the countryside. Tallahatchie Farms, among the last of the tenant resettlement projects of the Farm Security Administration, is still composed of small landholdings created in 1940 from three plantations (map 4.3). Most of the landholdings are still owned by blacks, but only a few actively farm them. The landholdings are cultivated by several farmers, known as *multitenants*, who create large operations by renting small landholdings from a number of owners (Prunty and Aiken 1972, 305–6). Today's farmers must achieve economies of scale to support large machinery inventories consisting of tractors, mechanical cotton harvesters, combines, and other expensive specialized equipment for plowing, planting, and applying fertilizer and herbicides. Such farmers who rent land are tenants, but they hardly resemble the earlier downtrodden ones.

Some of the landowners on Tallahatchie Farms still live in the modest frame houses whose architecture reflects their New Deal origin (fig. 4.2). But other owners have built new ranch-style brick dwellings. Also scattered across Tallahatchie Farms are small hamlets composed of hodgepodges of dwellings, including mobile homes and even old relocated tenant shacks. Most of the hamlets on Tallahatchie Farms consist of extended families with one- to five-acre lots for dwellings carved from the landholdings of relatives.

Near Tallahatchie Farms is Rainbow Plantation. The area surrounding it, including the municipality of Webb, a town with a population of 605 in 1990, illustrates other aspects of local black population redistribution, new settlement patterns, and the role of federal agencies (Aiken 1987, 1990). The em-

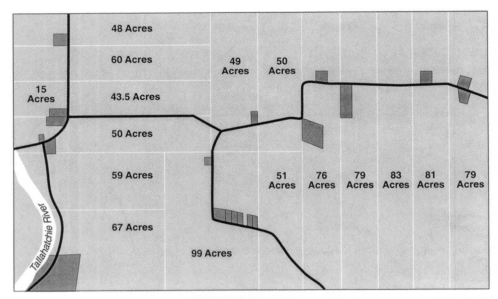

Map 4.3. The Mahoney area of Tallahatchie Farms, Mississippi, in 1987. Most land-holdings shown are not farmed by the black families who own them. Several multi-tenants lease these lands for farming. The small lots partitioned from the original land-holdings are typical of the new rural settlement pattern of blacks in the plantation South. *Source:* Tallahatchie County, Mississippi Tax Assessor's Office.

ployees of Rainbow Plantation live in a row of houses for machinery opera-tors, which replaced the tenant houses that were dispersed throughout the fields. Goose Pond subdivision between Rainbow Plantation and Webb has 85 single-family houses constructed in 1972 by the Farmers Home Administra-tion for home ownership (fig. 4.3). Although the agency does not discriminate racially, all the houses are owned by blacks, some of whom once were tenants on Rainbow Plantation. The monthly payments are federally subsidized for low-income households. Goose Pond is served by Webb's water and sewage systems, but the subdivision is not within the municipal limits. The location illustrates the new politics of race in the plantation regions. Webb is a white-dominated enclave in a county that has a black majority. If Goose Pond were in Webb, the town would be more than 50 percent black rather than 47 per-cent.

Churches

In addition to housing, African American churches reveal the commitment of blacks to the plantation regions. The black church was more than just an organization for worship, and its modest buildings masked its importance. From Reconstruction through the civil rights movement, most black leaders were clergymen. The African American church was the one place free of constant domination or interference by whites. Clifton Taulbert's testimonial is representative of the church's traditional central role: "It was closer to our hearts than our homes—the colored church. It was more than an institution, it was the very heartbeat of our lives. Our church was all our own, beyond the influence of whites, with its own societal structure" (1989, 91–92).

Fig. 4.2. An original Farm Security Administration house on Tallahatchie Farms, similar to scores of dwellings constructed by Franklin Roosevelt's New Deal agencies across rural America during the Great Depression. Although inexpensive, such houses were mansions to the small number of poor tenant farmers who qualified for a new beginning in agriculture as landowners. Photograph by Charles S. Aiken, May 1987.

Fig. 4.3. A street in the Goose Pond subdivision, Tallahatchie County, Mississippi. Photograph by Charles S. Aiken, May 1986.

Since the civil rights movement a significant part of black leadership lies outside the church, but churches are still the primary institution in the lives of most blacks in the plantation regions. Where only a few years ago black churches were modest frame structures, some in abysmal physical condition, today many are new air-conditioned brick buildings with Sunday school rooms attached to well-furnished auditoriums (fig. 4.4). Investment in church buildings perhaps more than in new dwellings reveals that blacks are committed to the plantation regions.

Conclusion

The reasons why plantation areas of the South were not true homelands for black people are numerous. As slaves blacks did not choose to live in these areas. And following their emancipation they lacked both the wherewithal and the education to leave the plantation areas. Sharecropping became a new form of servitude. Indeed, most blacks who left after about 1920 were forced to do so by collapsing plantation agriculture in some regions and agricultural mechanization in others. The exodus over the next forty years took them to urban areas in the North and West. Many tried to forget their roots in the South.

But several reasons can be given why the southern plantation areas are homelands—or more properly unique homelands—for blacks. Many blacks who stayed gradually became homeowners, a trend accelerated by the 1960s War on Poverty (fig. 4.5). And through home ownership they gained a measure of control. Between 1960 and 1990 blacks' occupational structure shifted from farming and personal services to manufacturing and professional services, and they slowly began to move into the middle class. Perhaps the most

Fig. 4.4. Dickerson CME Church in Fayette County, Tennessee, exemplifies new church buildings of blacks constructed since passage of the 1964 and 1968 Civil Rights Acts and the 1965 Voting Rights Act. Churches represent a new prosperity brought by general economic improvement in large areas of the rural South and the end of discrimination in hiring blacks for manufacturing and professional jobs. Photograph by Charles S. Aiken, June 1982.

Fig. 4.5. The home of a black rural nonfarm family in Tallahatchie County, Mississippi. By 1970 home ownership among blacks in the plantation regions was increasing rapidly. Photograph by Charles S. Aiken, 1987.

compelling reason for plantation areas as homelands, however, is that the children and grandchildren of those who left the South between 1920 and 1970 find part of their heritage in them. For many who left for the North and West, southern plantations meant only bad memories of poverty and discrimination. For blacks who remain and those who rediscover them, the plantation areas are the homes of their ancestors in which family and traditions are deemed to be good.

CHAPTER FIVE

The Creole Coast
Homeland to Substrate

· · · · · · · · · · · · · · · · · · · ·

Terry G. Jordan-Bychkov

Homelands, like all human undertakings, are not immutable. Change comes, though it differs from one homeland to another. Some evolve into independent nation-states, while others survive for millennia, yielding slowly and resistantly to the forces of assimilation. Still others—particularly in North America—perish after a relatively short lifespan. What happens when homelands die? Do they vanish without a trace, or does some residue of regionalism, some geographically discrete ethnic substrate, survive, allowing the former homeland to remain culturally distinctive?

I seek the answers to these questions in the maritime fringe of the American South, a region I call the *Creole Coast*. In the vernacular, parts of this region bear other names, such as *Low Country* and *Tidewater*, but these tend to mask a wider cultural-ecological unity that encompasses a coastal belt stretching from the Chesapeake Bay to the Texas Coastal Bend (map 5.1).

The Creole Coast as Homeland

In part, the special character of this littoral, this *place*, rests in its natural environment. Forming the low-lying, table-flat, poorly drained outer coastal plain of the South, the Creole Coast conforms well to Edwin H. Hammond's type A-1a landform (1964). In terms of climate, soils, flora, fauna, and disease ecology, the Creole Coast is more closely linked to the adjacent West Indian is-

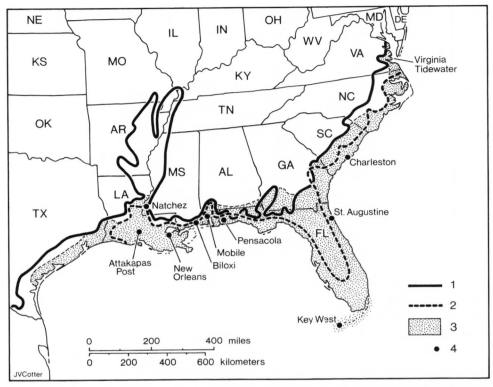

Map 5.1. The Creole Coast, 2000. Key to symbols: 1 = northern border of Edwin Hammond's type A-1a landform (table flat, low); 2 = border of Jay Edward's "Creole Coastal Culture Zone"; 3 = "Creole Coast," as used in this article; 4 = noteworthy, se-lected historical implantments of Caribbean creole culture, achieved by Cubans, His-paniolans, Jamaicans, Barbadians, and others. Courtesy of John V. Cotter.

land world than to the rest of the United States. Ecologically, the region is more Caribbean than North American, a fact recognized by any number of natural scientists (Ekman 1953).

More important, *people* also link the Creole Coast to the Caribbean. West Indian islands such as Barbados and Jamaica played important roles in the set-tlement of the Carolina Low Country and the Chesapeake Tidewater. His-paniola and Cuba mothered diverse Spanish and French colonies on the Cre-ole Coast, from St. Augustine west through Pensacola, Mobile, Natchez, and

New Orleans. These settlers bore a hybridized culture, the chief ingredients of which were Iberian, French, and African, with lesser or greater admixtures of English, Celtic, and Amerindian traits. Bloodlines were as mixed as the culture was creolized. The blending of these diverse peoples had occurred earlier, on the Caribbean islands, and the Creole Coast merely represented the final, outermost niche-filling by a tropical insular culture seeking environmentally compatible lands for expansion. Preadapted to this quasi-tropical littoral, the Caribbean folk quickly achieved a *bonding* with the land and the new colonies soon thrived.

In its first century and more of existence, the Creole Coast did not function as a homeland. Instead, it merely served as the rimland of the Caribbean culture area, a periphery and continental barrier to an insular core. A maritime people, they always looked seaward, toward the Indies, largely ignoring the interior of North America. In this interlude, well over a century in duration, the element of *time* strengthened the bonding of people and place along the Creole Coast.

Homeland status awaited the eventual independence of the United States in 1783 and the piecemeal American seizure of the remaining parts of the coast with the purchase of Louisiana in 1803, acquisition of the Floridas in 1819, and annexation of Texas in 1845. The independence of the United States and its subsequent policy of manifest destiny imposed a political boundary between the Creole Coast and the Caribbean islands, converting the long-seated coastal people into a cultural-ethnic minority. Timothy Flint, the well-known early geographer and a quintessential American, recognized the distinctive cultural character of the Creole Coast in the 1820s, observing in West Florida "a compound of Spanish, French, and American manners" (Flint 1828, 1: 471).

The Creole Coast homeland proved short-lived. The free peoples of the southern coastal plain sought secession from the United States—the Confederacy—but this attempt to seize *control*, essential to the longevity of homelands, failed. If we seek a date for the demise of the Creole Coast homeland, 1865 serves as well as any.

Does a legacy of regional cultural distinctiveness survive along the Creole Coast, or has it become simply another indistinguishable part of a larger nation-state? I suggest that the Creole Coast remains a region, a *place*, and that this represents the normal legacy of perished homelands. Fragments of the

Romano-Caribbean culture that formerly provided the foundation of the homeland remain abundant, lending a regionalism and sense of place to the littoral still today.

Romano-Caribbean Culture Fragments

In part, the enduring quality of Creole Coast culture derives from the fact that the traditional land-use systems prevalent along the southern littoral were of Caribbean origin. Two intertwined systems, or adaptive strategies, prevailed from the earliest colonial period—the slave plantation and the "cowpen" culture. Of these, the more important was the plantation. I favor Leo Waibel's notion that the slave plantation system sprang from the use of subsaharan African slaves in Moorish Europe (Waibel 1941, 156–60), a practice that survived the Christian reconquest in both Andalucía and southern Portugal (Pike 1967). Sugarcane spread with slavery from Moor to Latin Iberia. Before Columbus, the Portuguese established slave-based sugar plantations on the African islands of São Tomé and Príncipe in the Gulf of Guinea, and the system spread from there in the early 1500s to Brazil and the West Indies. Britons adopted this Latinized system and its staple crops in the West Indies, where Barbados developed as their prototypical island. Barbados borrowed heavily from Brazilian plantation culture, including purchases of slaves from the Pernambuco market, and as a result Portuguese influences entered the British West Indies (Beckles 1989, 117). The role of Barbados in the subsequent colonization of the British West Indies and the Carolina-Virginia Low Country is well known, and we must acknowledge the degree to which these planters became Latinized during their sojourn in the Caribbean. South Carolina's 1690 slavery legislation followed a Barbadian model.

Jamaica, seized by the English in the 1650s, became the principal arena of transfer of Spanish Romano-Caribbean influences into British Antillean culture, complementing the Portuguese infusions into Barbados. With the plantation system came its staple crops. The variety of tobacco raised on Chesapeake Tidewater plantations came from the Spanish Main, South Carolina's indigo apparently from Jamaica, both Sea Island cotton and the appropriately named "Creole" cotton from the West Indies, and Louisiana's sugarcane from the same source. Similarly, the early South Carolina practice of using Amerindian slaves, still much in evidence as late as 1710, merely echoed the older Spanish *encomienda* system of the Antilles (Carney 1993, 29; Morgan 1975, 90).

Complementing the plantations was a second major adaptive strategy along the Creole Coast. The cattle and hog herding system, or cowpen culture, long dominated the greater part of the Low Country, from Albemarle Sound to southeast Texas. It grew from multiple implants of Spanish Antillean ranching at various points along the Creole Coast (Jordan 1993, 78–84, 109–20; Otto 1986). Anglo-Jamaicans brought their hybridized version of the cowpen culture to South Carolina in the 1670s, and Antillean Spaniards in Florida and Haitian French on the central Gulf Coast also made enduring implantments. The West Indian Spanish roots of the cowpen culture are easily detected. Iberian longhorns remained the dominant cattle breed from southern Georgia to southeast Texas at least as late as the 1840s, including local variants such as the "Florida cow" (Mealor and Prunty 1976, 364–65). Similarly, the small, agile Andalusian pony, referred to as the "Seminole horse" in Florida in the 1770s, the "Spanish tacky" in the Mobile area in the 1820s, and the "Creole pony" in southwestern Louisiana in the 1850s, long remained common in the Low Country and perhaps provided bloodlines for the modern American quarter horse (Bartram 1791, 118–21; Olmsted 1857, 393; Flint 1828, 1: 490). The black razorback hog so important in the southern pine barrens and swamps apparently also sprang from Spanish West Indies stock and ultimately from Extremadura. Elsewhere I have proposed that the southern herder dog, variously called the Catahoula hound, brindle cur, and leopard dog, derives from feral descendants of the Spanish "dogs of the conquest," employed with great success and cruelty in the early Indies. Today, the remaining concentrations of this herding cur all lie in the Low Country, including the "Cracker" country of the Georgia-Florida-Alabama borderlands, the Mississippi floodplain of Louisiana, and the Big Thicket of southeast Texas (LeBon 1971; Varner and Varner 1983; Jordan 1993, 82–83, 119, 121).

Accompanying the diffusion of the plantation system and cowpen culture into the Creole Coast was a complex of West Indian foods and beverages linked to the Antillean Spaniards and French. Among these I would confidently list rum, pork, barbequed meat, chile peppers, rice, and, of course, Louisiana Creole cuisine. Southern dent corn also reputedly came from the Spanish Main, and I wonder whether the ubiquitous small grape arbors seen on Cracker farmsteads and in small-town dooryards of the Georgia Low

Country, as well as in St. Augustine, have a Mediterranean origin (Manucy 1992, 124–25). And what are we to make of the chile pepper, oddly called the *datil* (or literally "date") and grown still today in the St. Augustine area of Florida? Ethnobotanist Jean Andrews, who brought this variant to our attention, identified it as *Capsicum chinese* Jacquin, of West Indian rather than Mexican origin (fig. 5.1). It resembles outwardly a date, and hence its curious name, but inside tastes like sheer hellfire (Andrews 1993, 71–73; Andrews 1995). The datil remains the only *Capsicum chinese* to be cultivated commercially in the United States, if on a small scale and locally. At least four pepper sauces are bottled locally from the datil—Bill Wharton's "Liquid Summer," Oochies "Redneck," "Dat'l Do-it," and "Devil Drops." Other varieties of chile peppers provide the basis of numerous additional bottled sauces produced at various places along the Creole Coast, including New Orleans and Avery Island, Louisiana. These all reveal a culinary link to the West Indies. Grocery store shelves and menus always reveal cultural truths, if we will but notice.

People often lie, but their words, their vocabularies, never do. The English dialect spoken along the Creole Coast contains a sizable number of Romance loanwords and Caribbean usages that came by way of the West Indies. As in the case of the cowpen culture, multiple diffusions occurred, all along the seaboard. Anglo-Jamaicans introduced to South Carolina words such as *savanna* (an Amerindian word early adopted into Antillean Spanish), *palmetto*, *jerked* meat, *crawl* (hog farm, from Antillean Spanish *corral*), and probably *barbecue* (which I prefer to attribute to Caribbean buccaneer French *barbe-et-queue* (the whole animal, "beard-and-tail"). All of these loanwords occur in Jamaican English and must have reached Carolina from that island (Cassidy and LePage 1980). Barbadians arriving in South Carolina brought *verandah*, derived from Brazilian Portuguese, and likely other Luso-Romance words as well. Modern toponyms such as Savannah in Georgia and Hog Crawl Swamp in Jasper County, South Carolina, testify to the durability of these introductions. Even earlier, at the founding of Jamestown in Virginia, the colonists used the Spanish word *pallisadoe* to describe their defensive paling, a borrowing English pirates operating in the Caribbean may have accomplished.

Another lengthy list of Romance loanwords appear to have entered the Creole Coast by way of the Spanish and French settlements of West Florida— the coastal district between the Apalachicola and Mississippi rivers, with centers at Pensacola, Mobile, and Natchez. I believe this district served as the

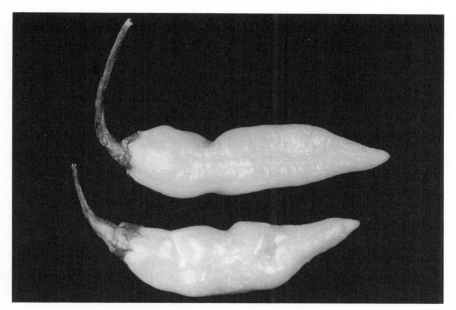

Fig. 5.1. "Datil" pepper (*Capisicum chinese* Jacquin) from the area of St. Augustine, Florida. This yellow chile pepper has a West Indian origin and is still cultivated in northeastern Florida. Photograph courtesy Dr. Jean Andrews.

greatest of all arenas of interaction between Caribbean Latins and Anglo-Americans, especially during the period of British rule between 1763 and 1781. To this region I attribute an array of Spanish loanwords that appear to have undergone French or some other modification before entering English. Many contain the *oo* sound, such as *calaboose* (jail, first documented usage in 1790s in Mobile and Natchez), *vamoose, galoot* (awkward person), *kiyoodle* (mongrel), *caboodle* (the whole lot), *maroon* (verb, to be stranded, from Spanish cimarrón, wild one), and possibly *saloon*. Other loanwords derived from West Florida may include *lingo, mosey, cavvyard* (group of saddle horses), and *cabras* (halter). Buccaneers of English and French derivation apparently developed a pidgin speech in the West Indies containing many of these Romance loanwords. Linguist J. L. Dillard has proposed that this Caribbean-derived, largely Romance "maritime lingua franca" was spoken among Caucasians and blacks along much of the Gulf Coast in the 1700s, fragments of which echo yet today in the regional vernacular (Dillard 1985, 137–42; Dillard 1987, 244–49; Dillard

1992, 117–29; Babington and Atwood 1961, 17). Listen, for example, to the modern speech of the Okenfenokee Swamp Crackers. You will hear words such as *calaberment* (loud noise made by animals), *trumpery* (cooking utensils and bedding), and *progue* (to prowl in the swamp)—and wonder about their origin (Harper and Presley 1981, 137–44). Dillard's lingua franca thesis is most compatible, spatially and temporally, to British and Spanish west Florida as the venue of most vigorous cultural-linguistic exchange between Creole and Cracker. The two cultures—destined to become substrate and overlay—met not only in the streets of Pensacola, Mobile, and Natchez, but also in places such as "the old Spanish Cowpen," shown up the Escambia River north of Pensacola on a 1771 map. Dillard uses the intriguing term "cowpen Spanish" (Dillard 1985, 141). Incidentally, this Romance pidgin should not be confused with "Mobilian," the Amerindian lingua franca of the Gulf Coast (Crawford 1978).

In addition, of course, many words later passed from Louisiana Creole French to English, to be carried westward along the Texas coast as far as Corpus Christi. Examples include *banquette* (sidewalk), *gallery* (porch), *lagniappe* (baker's dozen), *bayou, pirogue* (small boat), *couche-couche* (cornmeal dish), and *shivaree*. Louisiana Creole French, in fact, may serve as the best repository of vocabulary items of the old maritime lingua franca. The Okenfenokee Cracker use of both *chivaree* and *progue* (from pirogue?) suggests as much (Atwood 1962, 148, 161, 174–87; Babington and Atwood 1961, 8–9, 18–22; Harper and Presley 1981, 137, 143).

Reading the relic cultural landscape, a time-honored cultural geographical method, yields additional evidence of Romano-Caribbean influence along the Creole Coast. To this end, I can draw initially upon the innovative findings of Philippe Oszuscik and Jay Edwards. Oszuscik, a specialist in the folk architecture of the southern coast region, presented his findings on the diffusion of the "Tidewater raised" (or "Creole") cottage, a vernacular house type occurring along the coast, from Virginia to southeast Texas (fig. 5.2). The cottage is distinguished by full façade built-in porch development, by elevation on piers well above ground level, and by a classic tropical roof of two broad pitches broken, if at all, only by dormer windows (Oszuscik 1992b). Oszuscik also proposes a Caribbean origin for the hipped roof so common in Gulf Coast folk architecture (1992a, 163; 1994).

The Caribbean origin of the raised cottage with full porch enjoys general acceptance, but how and where this classic house type passed so vigorously into

Fig. 5.2. A "tidewater raised cottage"—a Caribbean folk house transplanted to the Creole Coast, here in the South Carolina Low Country. Of Caribbean origin are the elevation on piers and the full porch. The geographical distribution of this house type, also called the "Creole cottage," if ever compiled, would provide a good index to the extent of the Creole Coast culture. Photograph by Terry G. Jordan-Bychkov, 1980.

the English-speaking community of the Deep South remains far less clear. Oszuscik's findings seem to establish West Florida as that venue, or at the very least as one of several coastal zones of Caribbean architectural borrowing. He particularly favors Pensacola as the scene of interaction, and his emphasis upon West Florida dovetails nicely with Dillard's lingua franca thesis. Edwards convincingly proposes a multiple diffusion for the porch, including an early episode in lowland Carolina (1989, 45–48). Also likely of Caribbean architectural origin are the "raised courthouses" found on the southern coastal plain from the Carolinas to southeastern Texas (Fagg 1989).

Another reputed example of Caribbean architectural influence on the Creole Coast can be found in the method of wall construction known as *tabby*. Consisting of a concrete made from shells, lime, and sand, tabby first appeared in the St. Augustine area of Florida, where it became the most common method of construction (Manucy 1992, 32–36, 55–57, 68–77, 106, 115–16).

Tabby later diffused northward as far as the Carolinas, gaining its greatest acceptance in lowland Georgia, especially the Sea Islands. Individual specimens of tabby construction appear as far west as the Coastal Bend region of Texas (Murphree 1952, 391). Janet Gritzner, who wrote the definitive study of tabby, feels that Spaniards invented this construction method in the West Indies and modified it in Florida. The word probably derives from West African slaves, who in turn had obtained it in pre-Columbian times from Arabic (*tabbi*, "lime-earth mortar") or Ibero-Romance (*tapia*, "mud wall") sources. Direct lexical derivation from Spanish did not occur, because the Antillean Spaniards used other words for tabby, reserving *tapia* for rammed-earth construction (Gritzner 1978).

A final example of Caribbean architectural influence, and one which ultimately spread far inland from the Creole Coast, is the so-called *shotgun* house. John Vlach has convincingly demonstrated that this distinctive, long, deep house plan reached New Orleans from French Haiti at the time of the late eighteenth-century slave revolts (Vlach 1976). We need to learn, by the way, when and where the shotgun house passed from Francophone blacks to those speaking English (Jakle, Bastian, and Meyer 1989, 146).

No doubt many other traces of the Romano-Caribbean ethnic substrate remain to be detected. Archaeologist Thomas Loftfield, who excavated a Barbadian colony on the lower Cape Fear River in North Carolina, indicated as much in a recent paper, in which he pointed not only to pottery styles, but also to Caribbean influences in settlement patterns and defensive installations (1993, 72–73). In short, the cultural landscape, adaptive systems, foods, beverages, and vernacular speech along the Creole Coast reveal a broad-based ethnic substrate of Caribbean origin.

A Perished Homeland

If the Creole Coast is representative of perished homelands—and I strongly suspect it is—then we must conclude that a life after death awaits these special culture areas. In time, perhaps this regional or sectional afterglow will also disappear, but I believe otherwise. Even in the age of pervasive popular culture, people seem to seek regional identity, often achieving it by adopting elements from fading ethnic substrates. The Creole Coast leaves us the likes of Brunswick stew, Bourbon Street, barbecued pork, and much more.

CHAPTER SIX

Nouvelle Acadie
The Cajun Homeland

. . . .

Lawrence E. Estaville

Over the years, the idea that the Cajuns, Louisiana's Acadians, have been isolated from change has been advanced, reinforced, and disseminated to the American public and within the academic community. Contrary to this conventional view, it is proposed here, from an extensive search of the historical record and years of fieldwork, that dramatic change, at times terribly tragic, has filled the Cajun experience during the past four centuries.

During this time, émigrés from west-central France crossed the North Atlantic to become Acadians living on the far eastern margins of seventeenth-century Nouvelle France. Caught up in their tragic diaspora, *le grand dérangement*, many Acadians became Cajuns in the eighteenth century and created a new homeland, Nouvelle Acadie, in South Louisiana. In both places, Acadia and South Louisiana, these French-speaking people created homelands through time by impressing their culture traits onto the landscape and by modifying their ways of living to accommodate foreign physical environments and interaction with other peoples. In both places, these Francophones bonded to one another and to their lands, homelands that they not only came to control but to love, protect, and—ultimately—to lose. And in both places, invasions by Anglos caused the demise of these Gallic-derived homelands.

Frenchmen Become Acadians:
The Acadian Homeland in Nouvelle France

Early seventeenth-century French settlers of Acadia, present-day Nova Scotia, faced two important geographical changes. A harsh climate, which in January and February had temperatures 15 to 20 degrees lower than those of western France, forced the Acadians to undergo a wide range of adaptations in foods, clothing, and housing. Although the Acadians, after some difficulties, succeeded in cultivating their traditional crops of wheat, barley, and oats on the reclaimed saline marshes and in bringing the apple tree from France, they depended greatly upon a vibrant avenue of transculturation with the Micmac Indians that provided ways to supplement an initially precarious subsistence agriculture. The Acadians adopted Micmac techniques of hunting deer and moose and of fur trapping. They borrowed not only moose-pelt moccasins from the Indians but also tobacco and pipe-smoking (Clark 1968; Griffiths 1973; Brasseaux 1987).

The Acadians learned to make maple syrup and beer brewed with spruce buds, a noted anti-scurvy concoction preferred over the traditional apple cider. Most Acadians abandoned their French two-story, wood and masonry houses for the warmer, one- or two-room *poteaux-en-terre* cottage, a marriage of European design, Acadian building materials, and Indian construction and insulation techniques in which walls consisting of posts bound together by small branches and coated with a mixture of mud and clay sealed out the arctic blasts and supported a small attic and a European-styled roof thatched with reed and bark. All the while, their French vernacular language began to include Indian words to describe the new ways (Clark 1968; Griffiths 1973; Brasseaux 1987).

The second geographical problem was simply isolation. Acadia was neither in the mainstream of French colonization nor in the midst of a favorable artery of transportation, having been located on the shores of a relatively dead-end bay with unusually great tidal extremes. Thus, the seventeenth-century Acadians forged a unique blend of French and Indian folkways on an isolated edge of the North American frontier but in doing so became increasingly distant from their native French culture that continued on its own path of evolution (Clark 1968; Brasseaux 1987).

Although eighteenth-century Acadian neutrality and later recalcitrance were the strategies used to mitigate swirling geopolitical events, the Acadians, nevertheless, lost their homeland, fought off Anglicization in the American colonies as well as assimilation into a condescending society in France, and thus spent years in their *grand dérangement* searching for a Nouvelle Acadie—a land in which they again became pioneers. For many Acadians, the tragic diaspora came to a close in subtropical South Louisiana far from their frozen Canadian homeland (Winzerling 1955; Brasseaux 1987; LeBlanc 1962).

Creation of the Cajun Homeland: Nouvelle Acadie

Between 1765 and 1785 most of the 4,000 Acadians who ended their diaspora in present-day South Louisiana settled on the backslopes of the lower Mississippi River and its distributaries. Spurned by their own French monarch, about 1,600 Acadians, the largest single group, were transported by the Spanish crown to Louisiana in 1785 to try to stabilize Spain's economically unprofitable colony (LeBlanc 1962; Davis 1965; Comeaux 1992) (map 6.1).

Throughout the late eighteenth century, the Acadians, contrary to current automobile-society images of the Atchafalaya Swamp as an impenetrable watery barrier infested with reptiles, saw the basin as a highly porous landscape that allowed them to roam widely within it, to settle its western edge along Bayou Teche (displacing such native peoples as the Houma, Opelousa, and Attakapas), and to maintain constant contact between the Mississippi and the Teche. Indeed, the lower Mississippi River, its distributaries, and its Atchafalaya Swamp form a natural system of waterways perhaps finer than any other on the continent that ensured ease of early Cajun transportation (Davis 1965; Kniffen 1968; Estaville 1987, 1988) (map 6.2).

Cultural change was pervasive in South Louisiana. The Acadians abandoned wheat, barley, and oats for Indian corn, oriental rice, and West African okra. Flax would not grow in the hot, wet climate, so cotton fields and pre-Whitney gins began to dot the countryside. Broadcasting seed no longer worked in soil that needed furrows. Apples and cider vanished. Figs, peaches, and wine from vines of concord, white, and muscadine grapes were enjoyed. Agricultural practices thus radically changed as the Cajuns plowed rows for new crops that took advantage of a far longer growing season, fought off seemingly incessant and infinitely varied hordes of insects, and, perhaps most important, became slaveholders, thereby deserting a proud heritage of egalitarianism (Clark 1968; Brasseaux 1987; Foret 1980).

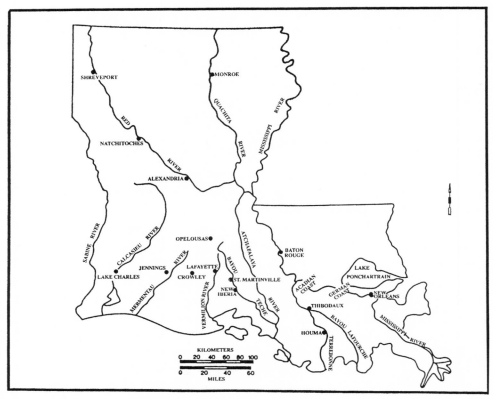

Map 6.1. Louisiana, 2000.

Sheep flocks shrank significantly at the expense of burgeoning cattle ranches—the huge *vacheries*, an industry learned from the Spanish. Barn size likewise shrank, for cattle no longer needed large winter refuges. The number of horses increased exponentially as the animal became central to agrarian Cajun life for both work and pleasure. Corn bread replaced whole wheat, sugarcane replaced the maple tree for syrup, and filé gumbo replaced *soupe de la toussaint.* French Creole (non-Cajun French born in Louisiana), Indian, and African cooking methods, particularly the red sauces introduced by West African slaves, crept into Cajun cuisine (Clark 1968; Comeaux 1992, 1996; Martin 1976; Post 1957, 1962; Brasseaux 1987).

Louisiana's subtropical climate forced Cajuns to change their entire wardrobe to loose-fitting, cotton clothing. The Canadian woolen shawl became the Louisiana cotton *garde-soleil*, a headdress with a wide, rigid brim that provided protection from the sun. Moccasins were worn in the short winter but were forsaken altogether during the remainder of the year.

The thick-walled, heavily insulated *poteaux-en-terre* cottage was not only unbearable in Louisiana's heat, but the high water table rotted the structure's wooden-post foundation and termites eagerly ate away the wood at ground level. The old Acadian cottage was quickly abandoned for what has come to be known as the Cajun house, a simple structure raised off the ground by cypress blocks and having large doors, matched windows and a front gallery for increased ventilative cooling, and cypress shingles called *merrain* covering a steep, gabled roof for better nocturnal heat radiation from the attic (fig. 6.1).

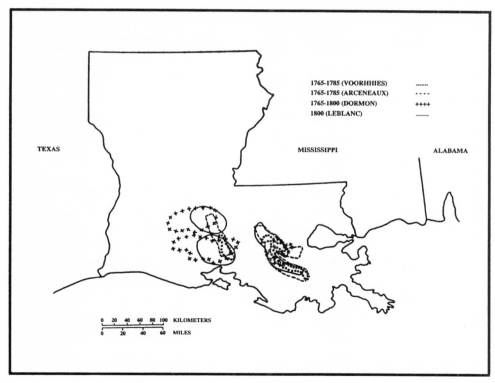

Map 6.2. Renderings of the Cajun homeland, 1765–1800.

Fig. 6.1. A Cajun house in Acadiana Village, Lafayette Parish, Louisiana. The Cajun house symbolizes the geographical experience of the Cajuns. When they arrived in their Nouvelle Acadie, the Cajuns underwent many cultural changes as they began to build and bond to their new homeland in south Louisiana. Today, however, like the Cajun homeland, the Cajun house is only a relic on the landscape. Photograph by Lawrence E. Estaville, April 1998.

The Cajuns borrowed this architecture from Frenchmen who had preceded them in settling Louisiana and who had brought the prototype from the West Indies. *Pieux*, roughly hewed cypress boards, and *bousillage*, nogging of mud and moss, were the main construction materials for the Cajun house, which, except for its painted front, weathered to a dull grey color (Brasseaux 1987; Post 1962; Comeaux 1992, 1996; Heck 1978; Robison 1975).

In the seventeenth and eighteenth centuries, geographical relocation into alien physical environments caused the French settlers to make striking cultural adaptations and borrow ways of living from Indians, French Creoles, Spaniards, and African slaves to survive in Acadia and South Louisiana. In the nineteenth and twentieth centuries, although the Cajun homeland would become geographically stable, invasions of Anglo-Saxon Americans by the thousands and new, exciting technologies would profoundly change Cajun life.

Anglo Invasion of the Cajun Homeland

Historic events filled nineteenth-century Louisiana—the Louisiana Purchase, statehood, the Civil War, and Reconstruction—and new technologies swept the state—steamboats, railroads, telegraphs, agricultural mechanization, electricity, and telephones. And the nineteenth century was one of critical, tumultuous cultural transformation in which an aggressive Anglo-Saxon nation displaced the French as Louisiana's predominant people.

Only a decade after the famous *New Orleans* arrived in its patron city did the steamboat age burst throughout the bayous of South Louisiana. The Attakapas Steamboat Company, chartered in 1821, served Bayou Teche, a slow-moving stream that snakes through the heart of the Cajun homeland west of the Atchafalaya Swamp, and in 1826 the steamer *Louisville* was the first to reach St. Martinville, the most important Cajun town on Bayou Teche. The Opelousas Steamboat Company in the 1820s provided transportation from Bayou Plaquemine through the Atchafalaya Swamp to a landing several miles north of Opelousas. During the latter half of the decade, steamboats began to ply the waters of Bayou Lafourche. By 1850 four boats regularly steamed the two days from New Orleans up Bayou Teche delivering and boarding freight and passengers at Franklin, New Iberia, and St. Martinville and stopping for commerce at all intermediate landings (Davis 1965, 1968; Kniffen 1968; Sandoz 1925; Comeaux 1972; De Grummond 1949).

After the Civil War, schooners again sailed up the Lafourche, Teche, Vermilion, Mermentau, and Calcasieu to engage in lucrative extraregional commerce based on South Louisiana's agricultural and timber production. Again the ubiquitous steamboat skimmed into the most remote locations for commerce and industry. Farmers shipped tons of agricultural commodities to market from "Louisiana's garden" along Bayou Teche; customized steamboats hauled three to four hundred head of prairie cattle at a time to New Orleans auction lots; and swampers towed huge rafts of cypress logs out of the Atchafalaya Swamp. In return, the popular *caboteurs*, the trading boats, again steamed throughout the bayou country, though this water-borne commerce had become much more specialized: fish, fruit, and ice boats; restaurant boats, saloon boats, and showboats with musicians, dancers, magicians, gamblers, zoos, and later even silent movies; boats with photographers, doctors, and dentists; and, of course, the store boats packed from stem to stern with groceries, dry goods, hardware, notions, and other merchandise. Indeed, the Cajun *petits habitants* of the easily traveled bayou country were far less isolated than the Anglo yeoman farmers trapped behind river rapids in the North Louisiana hills (Davis 1965, 1968; Taylor 1976; Kniffen 1968; Case 1973; Post 1957, 1962; Estaville 1988).

As early as the 1830s railroads began to track across the South Louisiana countryside, and the Gallic community enthusiastically supported rail transportation. The railway revolution was also a communications revolution. Mail between New Orleans and Franklin on the lower Tèche was delivered every third day in 1848. During the next decade, the mail runs of the New Orleans, Opelousas and Great Western, the state's second longest railroad, made it possible for Franklin's residents to read a New Orleans morning newspaper the same evening. But perhaps of greater significance was that the railroad introduced antebellum Cajun Louisiana to the precursor of today's electronic age—the telegraph (Reed 1966; Davis 1965; De Grummond 1949; Lathrop 1960; Carleton 1948).

The Louisiana rail net expanded substantially from less than 400 miles in 1860 to more than 2,000 in 1900, and 1883 marked the first through-train service between New Orleans and San Francisco, certainly an advantageous link that traversed the breadth of the Cajun homeland. Railways tied together South Louisiana's urban centers with the rest of the nation by century's end and they became the raison d'être for several bustling small towns in the Cajun homeland—Crowley, Jennings, and Rayne, for example (Estaville 1987, 1989; Davis 1965; Millet 1964).

Before the Civil War, South Louisianians thus felt the pervasive effects of transportation revolutions that rapidly pierced whatever barriers of geographical isolation that remained. Cajun communities were no exception; they too were transformed by the steamboat and the railroad. Easy transportation throughout South Louisiana surely encouraged extensive Anglo settlement in *Nouvelle Acadie*, thereby changing its landscape and its culture.

In the early nineteenth-century Cajun homeland, as in every southern region, there existed a complex mix of white and black, free and slave, rich and poor, educated and ignorant, mansions and shacks. An increasingly important part of this cultural milieu was the thousands of Anglo-Americans who began to stream into Louisiana even before the Louisiana Purchase and who deposited their English-speaking slaves deep in the Cajun homeland. During the century's first decades an intense Franco-American cultural clash developed. But by the 1840s Anglo economic and political hegemonies controlled Louisiana and had irreversibly begun to change the Cajun social fabric. American encroachment, in which Anglos bought up and consolidated fragmented, uneconomical French long lots, forced some Cajuns, mainly from the old "Acadian Coast" along the lower Mississippi in Ascension and St. James parishes and from the upper reaches of Bayou Lafourche in Assumption Parish, to move westward in what some scholars have termed the "second expulsion" and rapidly Anglicized those who remained (U.S. Population Schedules 1820, 1860, 1900; Arceneaux 1981; Rushton 1979; Comeaux 1972).

The Demise of the Cajun Homeland

By the outbreak of the Civil War, Anglo ways had greatly diluted the French culture. The two ethnic groups possessed similar occupational structures and personal and real wealth, had similar living conditions, and shared a prevailing illiteracy. The war itself became a tragic catalyst that intensified Anglo-French contact. Thousands of Union soldiers invaded the Cajun homeland and occupied it for more than a dozen years. The northern victory indelibly changed the lives of all Louisianians (U.S. Population Schedules 1860; Lathrop 1960; Winters 1963; Davis 1965; Edmonds 1979).

Anglos had also won the battle for state political hegemony. At the time of the Louisiana Purchase in 1803, there were roughly seven Frenchmen for every one Anglo-American in Louisiana, and at least a three to one ratio at statehood in 1812. By the Civil War 70 percent of Louisiana's population was Anglo. Fully appreciating the implications of the swelling Anglo ranks, the Gallic leadership secured a majority of delegates at the state's first constitutional convention and contrived its promulgation to ensure long-term French political control. Out of the Constitution of 1812 came the "French Ascendancy" of the 1820s. This Gallic political dominance was short-lived, however. After years of howling for political equity, Anglos captured control of the state government through their majority in the Constitutional Convention of 1845. The gubernatorial election of Cajun Paul Hebert in 1853 was the last breath of French domination in Louisiana (Tregle 1954, 1972; Howard 1971; Shugg 1936; Estaville 1984). Political scientist Perry H. Howard (1971, 227) calculated: "Of the thirteen elected governors before 1860, three were of French background, two were Acadians, and one had a French mother." No French governor was elected in postbellum Louisiana.

Anglos unequivocally won the struggle for state political supremacy, and they aggressively grabbed political power at every governmental level in the Cajun homeland. From the Civil War to century's end, the number of Cajun sheriffs was almost halved from five to three. Totals for 1900 census enumerators, grass-roots positions filled through political patronage, demonstrated the same pattern. In the "Acadian Coast" parishes of Ascension and St. James, only one of the 19 census enumerators was Cajun; and in Lafayette, St. Martin, and Vermilion parishes, all heavily populated by Cajuns, just 11 of 29 enumerators were Cajuns. On the other hand, of the town of Lafayette's first 16 mayors, 11 were Anglos. Seven of New Iberia's first 10 mayors between 1876 and 1900 were Anglos. In Ascension Parish, 12 of its 19 postbellum state senators were Anglos, as were 15 of its 24 state representatives, 10 of its 15 clerks of court, and 7 of its 13 coroners (U.S. Population Schedules 1860, 1900; Estaville 1984, 1987).

Most nineteenth-century Cajuns were farmers who supplemented their crops and livestock with the bounty of forest and stream, and during the century they borrowed Anglo agricultural practices. André LeBlanc, a Cajun from Assumption Parish, expressed his appreciation in an 1850 article in the renowned *DeBow's Review* (Comeaux 1972, 10): "We owe in great measure, to the inhabitants of the Carolinas and Virginia, who have settled among us, the great improvements we have so far made in agriculture." In the last two decades of the century, hundreds of midwestern families, attracted by familiar flat grasslands at cheap prices in a subtropical climate, swept onto Louisiana's southwest prairies. Rice agriculture alone, with its 26-fold postbellum increase, its modern mechanical methods, and its midwestern progenitors, markedly changed southwest Louisiana and the many Cajuns drawn into the center of the rice industry. And because they worked with newly invented steam plows, Osborn harvesters, Randolph rice headers, Engellery hullers and polishers, and centrifugal pumps used within extensive canal irrigation systems, these Cajuns became far more advanced in agricultural technology than the small Anglo cotton farmers of North Louisiana or other areas of the South (Ginn 1940; Millet 1964; Post 1962; U.S. Department of the Interior 1864, 1902; Comeaux 1978; Davis 1965; Kniffen 1968; Estaville 1984).

South Louisiana towns, like those throughout the United States, were caught up in a wave of late nineteenth-century innovation. Three years after Alexander Graham Bell invented the telephone in 1876, a "talking telegraph" company was incorporated in New Orleans. Just two years later Donald-sonville, 80 miles up the Mississippi into the Cajun region, inaugurated a local telephone exchange that linked the town to New Orleans in 1883. Ten years later construction began on an ambitious telephone system that by 1900 connected most towns and villages in or near the Teche country: Jeanerette, Olivier, New Iberia, Loreauville, St. Martinville, Lafayette, Breaux Bridge, Arnaudville, Sunset, Grand Coteau, Opelousas, Washington, and others. In 1887 New Orleans installed its first electric lights. Within little more than a decade New Iberia, Lafayette, Crowley, and Opelousas did likewise. In 1902 a New Iberia physician purchased one of the first automobiles in Louisiana. In 1909 Donaldsonville began to enforce automobile speed ordinances (Estaville 1987; Davis 1965; Millet 1964). Of course, if most Cajuns could not afford such new-fangled things as telephones, electric lights, or automobiles, they were no different from most Anglo farmers in North Louisiana, throughout the South, or in other parts of the nation. Yet, the essential point here is that neither geographical nor social barriers prevented the diffusion of the latest technological innovations throughout the towns and villages of the Cajun homeland and that these inventions began to affect the Cajun culture, not after World War II but before the turn of the twentieth century. Indeed, times of swirling innovation—mechanized agriculture, oil and sulphur booms, electricity, and telephones—badly blurred the most sacrosanct icons of the Cajuns during the last decades of the nineteenth century (Estaville 1984, 1987) (fig. 6.2).

Like the inventions that scintillated through their streets, the "bright lights" of Cajun towns began to beckon to rural folk to forsake their bucolic life-style. Contrary to those who have characterized Cajuns as only simple, rural folk, nine of the state's 14 urban places outside New Orleans in 1900 were located in Cajun Louisiana. Cajuns comprised almost 15 percent of New Iberia's white population, the region's largest urban concentration, one-fifth of Houma's, about a quarter of Lafayette's, and nearly a third of Thibodaux's (Calhoun 1979; U.S. Population Schedules 1900; Estaville 1987, 1988).

But the most significant change in Cajun culture in the nineteenth century was the tremendous erosion of its language. The demise of French political influence critically affected the essence of the Cajun culture, its language. As early as the Reconstruction Constitution of 1864, the Louisiana legislature

Fig. 6.2. Modern-day rice dryer in southwest Louisiana. With its new mechanical methods, the expanding commercial rice industry on Louisiana's southwest prairies drew the Cajuns into rice farming. Photograph by Lawrence E. Estaville, October 1974.

stridently required that the "general exercises in the public schools were to be henceforth conducted in the English language" (Oukada 1978, 20), and four years later the infamous "Carpetbagger Constitution" categorically declared that "no law shall . . . be issued in any other language than the English language" (Oukada 1978, 9). Such arrogant legislation underscored the statutory support of the longstanding Anglo attitude that language predominance would melt the Cajuns into southern society. Frederick Law Olmsted (1953, 319), for instance, perceived this Anglo-Saxon ethnocentrism during his antebellum travels through Cajun Louisiana: "The Americans would not take the trouble to learn French."

Surprisingly, the Catholic Church, the venerable Gallic institution for religious and secular education, was an influential agent in inculcating American values in the Cajuns. For example, Gabriel Audisio (1988) investigated the dynamics of the French-English language conflict within St. Joseph's Catholic

Church in antebellum Baton Rouge. Although during the 1850s half of its 292 families were French (56% Creole, 44% Cajun) and an overrepresented number of French wardens controlled the parish council, both St. Joseph's priest and its parishioners had been abandoning the French language throughout the 1840s. Father M. Brogard, St. Joseph's priest, in explaining why he no longer wanted to preach in French, complained to his bishop in 1843: "Your highness probably wants to know if the remarkable crowd present on the morning I preached in English was not declining when I began to preach again in French? No *Monseigneur* it did not decline, *it quite disappeared*" (Audisio 1988, 360, Brogard's emphasis).

According to Cécyle Trépanier (1986), by 1917, 12 of her 35 sampled communities (34%) in French Louisiana had never had Catholic church services in French or such services had been discontinued in the late nineteenth or early twentieth century, a trend that would see another 12 communities discontinue French church sermons before the end of World War II. In fact, by 1980 only one of these 35 communities had French church services on a regular basis—one Sunday each month.

The two most pivotal reasons for adopting English as the language of Catholic church services, particularly in the last decades of the nineteenth century, were: (1) to gain national acceptance, the Catholic church hierarchy in French Louisiana was determined to Americanize its members; and (2) many priests as well as parishioners could not speak French. Such an English language strategy was also aimed at the strongly Roman Catholic Cajun homeland, where Protestant missionaries had been doubling their congregations since the antebellum period (Trépanier 1986; Estaville 1984).

Not only did the use of French erode significantly during the nineteenth century, but the language itself changed markedly. The unique patois of South Louisiana evolved from different "gumbos," each based on a French "roux" of several languages. The introduction of English, however, had the greatest effect in corrupting the French spoken in the Cajun homeland. When listing 147 English words and meanings that in some way had crept into spoken Cajun French by 1939, Harley Smith and Hosea Phillips (1939) blamed this historical corruption on an educational process that was permeated with English words, ideas, and viewpoints and that had restricted contact with French culture, especially French literature. Carl A. Brasseaux certified (1978, 212) that "Cajuns became the target of constant pressure by Anglo-Americans to accept

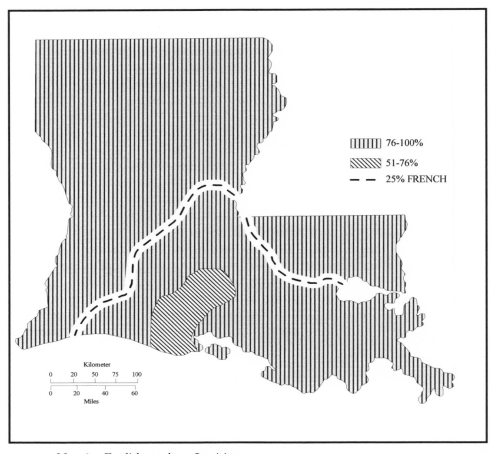

Map 6.3. English speakers, Louisiana, 1900.

an educational system made in their own image and the values which had produced that system."

These were not new thoughts, though. Alcée Fortier (1891, 85), celebrated Creole member of Louisiana's literati, lamented when he visited some rural Cajuns in 1891: "Education will, of course, destroy their dialect, so that the work of studying their peculiar customs and language must not be delayed long." Such anxiety was well founded. In 1900 more than three-quarters of the residents of all but three Louisiana parishes claimed they could speak English (map 6.3). Only in St. Martin (60% English-speaking), Lafayette (61%), and

Vermilion (62%) parishes did more than a quarter of the population tenaciously refuse to speak anything other than their local French patois (U.S.
Populations Schedules 1900; Estaville 1984, 1987, 1990).

Moreover, many Cajuns who struggled to maintain their cultural heritage
at least through their distinctive fiddle and accordion music, if not their language, were unaware that their forefathers had translated Anglo-Saxon songs
possessing a foreign pentatonic scale directly into French, at times with little
modification that left words that do not rhyme: "Paper of Pins" became "Un
papier d'epingles," and "Billy Boy," "Billy Garçon" (Brandon 1972; Oster
1959; Estaville 1987). It should not be surprising, then, that by the turn of the
century Cajuns were drinking beer, playing baseball, and speaking English
with their fellow Americans (Estaville 1990).

Meanwhile, the Cajuns expanded geographically out from the "Acadian
Coast" and bayous Lafourche and Teche across much of South Louisiana, after which they began to drift westward: from the east, pushed by an Anglo ap-
petite for rich delta soil, a billow of Italians that broke across the South
Louisiana "Sugar Bowl," and an obsolete arpent survey system that left on the
land splinters of uneconomical long lots; and from the west pulled by cheap
prairie land, by commercial rice agriculture that midwestern families nurtured
along the railroad tracks of southwest Louisiana, and later by a burgeoning
East Texas petroleum industry (Rushton 1979; Comeaux 1972; Scarpaci 1972;
Taylor 1950; Millet 1964; Post 1962; Louder and LeBlanc 1979; Estaville
1986).

The Cajun Homeland in the Twentieth Century

In the twentieth century freedom brought by the automobile and information
beamed by radio, movies, and television captured the Cajuns and fired the Gallic cultural dissolution. Epitomizing such disintegration, *L'Abeille*, Louisiana's
last important French-language newspaper, suspended publication in 1923,
following the moribund path of 26 other French-language and 21 bilingual
newspapers that had ceased publishing between 1860 and 1900 (Parenton
1949; Saucier 1951; Brasseaux 1978; Tinker 1933; Oukada 1978).

In 1901 oil struck near Jennings marked the beginning of an "oil boom"
that became an economic magnet for thousands of Anglos who settled in the
Cajun homeland in the twentieth century. Later, the construction of huge pe-

troleum refineries at Baton Rouge, Lake Charles, and Port Arthur, Texas, lured many better-educated Cajuns to higher-paying jobs (fig. 6.3). These Cajun plant workers at first commuted by automobile and then moved closer to the refineries, particularly those in East Texas. However, at mid-century when the famous "Oil Center" was constructed in Lafayette in the heart of the Cajun homeland, Anglo professionals joined gangs of construction workers and "roughnecks" who had come by the hundreds to service the forest of offshore oil rigs being erected in the Gulf of Mexico. Meanwhile, Cajun families felt the personal agonies of both world wars, took advantage of continued educational opportunities, and traveled throughout the nation via railroads, superhighways, and airlines (Kniffen 1968; Davis 1965; Louder and LeBlanc 1979; Estaville 1987, 1993) (map 6.4).

Today the more than half million people of Cajun ancestry comprise about 12 percent of Louisiana's population. Cajuns are still married to their auto-

Fig. 6.3. Petrochemical refinery in Lake Charles, Louisiana. Soon after oil was discovered in Louisiana in 1901, Cajuns began to work in the oil fields and later drove or moved to work at nearby oil refineries. Photograph by Lawrence E. Estaville, 1974.

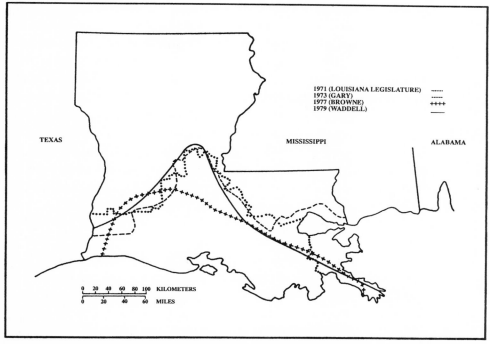

Map 6.4. Renderings of the Cajun homeland, 1971–79.

mobiles, watch CNN and MTV on cable television, hold critical state political clout called "Cajun Power," live in three-bedroom, two-bath brick homes,
program computers, use their cellular phones and electronic mail to commu-
nicate globally, have been swept up in the current "gambling craze," and,
though their spicy foods remain hot commercial "properties" that have won
the nation, eat more Big Macs than *boudin*.

Cajuns have thus become mainstream urban Americans. Yet recently, because of the efforts of the Council for the Development of French in Louisiana
(CODOFIL) to revive the French language and because of a propitious national spotlight, a resurgent pride in their Cajun heritage has seized their
South Louisiana homeland. "Rajin' Cajun" athletes of the University of
Southwestern Louisiana (now LA-Lafayette) have gained national promi-
nence. From the small prairie town of Crowley have come a colorful Cajun
governor, Edwin Edwards, and his protégé John Breaux, who is now a senior

U.S. senator. Reflecting this Cajun revitalization are French festivals—from the simple *boucherie* to the ostentatious Mardi Gras—that seem to have become more effervescent, filled with the esprit of *laissez les bons rouler* (let the good times roll) (Estaville 1993).

It seems that today America wants to know more about the Cajun homeland, a region that has been dramatically portrayed either as sinister swamps filled with reptiles or picturesque bayous draped with moss and magnolias. Centered in Lafayette, the nascent Cajun renaissance still faces the test of time. Yet the ironies are many. As the Cajuns anxiously try to regain their identity from the secure embrace of a national culture, myopic perceptions of their heritage become media events.

Conclusion

French settlers in a relatively remote part of the New World became Acadians; forced to flee their Canadian homeland, Acadians became Cajuns in the subtropical wetlands of South Louisiana; and thousands of ethnocentric Anglo-Saxon invaders saw to it that the Cajuns would become Americans. Compared with any other group of Europeans in North America over the years, these French people were forced to make the most radical changes, including the loss of their language, the essence of a culture. Although French Creoles and Anglos scorned the word Cajun and the *joie de vivre* of "Cajuness" for more than two centuries, in the last decades of the twentieth century a newly discovered Cajun pride has mesmerized South Louisiana. Yet, today the Cajuns' piquant cuisine and vibrant music are more commercial products than cultural cement.

Likewise, the Cajun homeland has changed remarkably since its eighteenth-century creation. A relic folk landscape remains faintly scattered here and there, and the Cajuns continue to defend the sanctity of their homeland as if they were still in control. But because of the Anglo domination of South Louisiana, first through sheer numbers, then via economic and political power, the Cajun homeland today is more myth than reality.

CHAPTER SEVEN

La Tierra Tejana

A South Texas Homeland

· · · · · · · · · · · · · · · · · · ·

Daniel D. Arreola

South Texas is the Tejano (Texas Mexican) homeland. This region is part of a Texas-Mexican rimland where contiguous counties have populations that are more than half Mexican American. Thirty-three counties make up the homeland, a borderland subregion with 1.7 million people, 71 percent of whom are of Mexican origin (map 7.1).

Complementing this demographic dominance are several environmental, social, and political characteristics that distinguish this subregion in the Hispanic American borderland, a zone that stretches from Texas to California. South Texas is the only coastal, semiarid subtropical lowland that is bounded by a riverine habitat. This environment has both accommodated and stamped South Texas Mexicans with a special sense of place, what I call la Tierra Tejana, the Texas Mexican land. A long settlement legacy in this region—nearly three centuries—results in a strong attachment to place evident, for example, in the many small towns with Mexican landscape attributes like plazas (fig. 7.1). South Texas has also been a cradle of ethnic political ferment and rebellion. Control in the region is now largely vested in Hispanic elected officials, more so than in any other borderland subregion.

In this chapter I examine place bonding between Tejanos and their *tierra*. The analysis proceeds through an assessment of people, place, time, and control as critical variables in the formation of a homeland. Then, through two

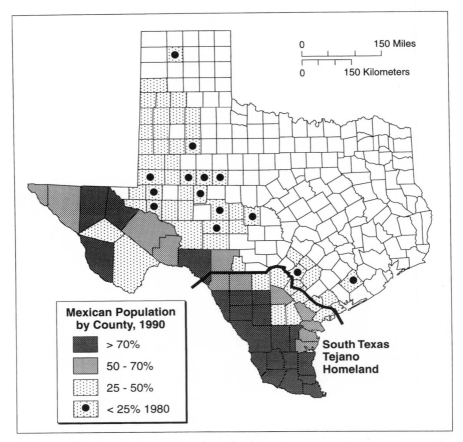

Map 7.1. The South Texas Tejano homeland (2000) encompasses some 33 counties within the horn-shaped outline created by the Rio Grande and Gulf Coast below the Balcones Escarpment and south and west of the San Antonio River. *Source:* U.S. Bureau of the Census, 1990a.

case studies at different scales—a small town, San Ygnacio, and a metropolitan area, San Antonio—I examine the Tejano homeland's place personality.

Tejanos

In 2000 California and Texas were home to 13.4 million of the nation's almost 20.6 million people of Mexican heritage. The Golden State counted 3.4 million more Mexican Americans than Texas, but California's rank as the state with the greatest number has only recently been achieved (U.S. Bureau of the Census, 2000). Before 1950 Texas was the undisputed demographic leader of Mexican Americans in the United States (Boswell 1979). Although the roots of the *Californio* (California Mexican) subculture are chiefly in the nineteenth century, the lure of a postwar economy drew thousands of borderland Mexican Americans especially from Texas and Arizona to the rapidly growing cities of Southern California. In the same era, agricultural development in the rural counties of the San Joaquin Valley attracted Mexican American migrants from Texas (Arreola 2000).

Despite the growth of the Mexican American population in California, today only one county—Imperial, in the southeast corner of the state—counts greater than half of its population as Mexican in origin (map 7.2). In neighboring Arizona and New Mexico the number of counties that are predomi-

Fig. 7.1. Plaza del Pueblo in San Ygnacio. In 1874 villagers laid out a gridiron plat and central plaza in San Ygnacio, today a community of a thousand people located on the Rio Grande 35 miles below Laredo. The many plazas found in communities in La Tierra Tejana serve as focal points for residents. Photograph by Daniel D. Arreola, 1991.

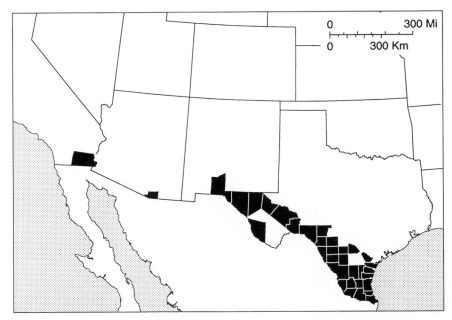

Map 7.2. Borderland counties with populations that are more than half Mexican in origin, 1990. *Source:* U.S. Bureau of the Census, 1990b.

nantly Mexican in origin is similarly few: Santa Cruz and Doña Ana, respectively. Along the Texas border, by comparison, no less than 32 counties are greater than half Mexican, and 20 are more than 70 percent Tejano Mexican ancestry; 17 of these counties are part of the South Texas homeland (map 7.1) (Arreola 1993b). This proportional concentration of a single Hispanic subgroup across so vast a geographic area may make the South Texas borderland the largest ethnic subregion of the United States, a veritable Tejano homeland along our south-central border (Arreola 1995a).

This distinctive borderland cultural geography exists, in part, because of the attenuated boundary along this Rio Grande periphery and the comparatively shorter lengths of border in states to the west of Texas. Nevertheless, boundary lines alone cannot explain this unusual geographic concentration. The historical inertia of Mexican settlement in South Texas and the persistent interactions over time between populations on each side of the borderline—northeast Mexico and South Texas—create a prominent subcultural area along the Rio Grande (Paredes 1993). Some scholars have argued that this border is imaginary and that both sides have more commonalities than differences, yet it must be acknowledged that *Tejano* as a cultural designation is only applied to people and things east and north of the river. The term denotes "a Texan of Mexican descent, thus a Mexican Texan or a Texas Mexican" (Benavides 1996, 238). *Tejano* is so widely used today that it is considered a naturalized word in the Texas lexicon.

There is little argument that the northeast Mexico periphery—especially the states of Nuevo León, Tamaulipas, and Coahuila—is a source area for Mexican American South Texas, but the two subregions are separate national and sociocultural entities and have been so for almost a century and a half (Arreola 1993c). Tejano roots in northeast Mexico are visible through a number of cultural signatures, some of which I elaborate on later. Two additional important people characteristics are a distinctive pattern of Tejano surnames and a historical link to Sephardic Judaism in the region.

Robert Buechley has studied surnames among borderland Hispanic American subgroups, and Richard Nostrand (1992a, 8–9) focuses on surnames as evidence of Hispano distinctiveness. Particular surnames like *Apodaca, Baca, Gallegos, Luján, Salazar* and more than 30 others are said to be more common in New Mexico, and, therefore, may be diagnostic of the diffusion of New Mexican Hispanos to other borderland states like California. A preliminary investigation that compares common Spanish surnames in San Antonio and Los Angeles suggests that Tejano distinctiveness may be equally valid. I assessed surnames in telephone directories for greater San Antonio (1994) and for greater Los Angeles (1991). Hand tabulations made of Spanish surname entries from each directory reveal specific surnames that are much more common in the Texas city (table 7.1). Whereas some names appear to be only moderately at variance, others like *Garza, Treviño, Villareal,* and *Cantú* may be particularly Tejano surname markers.

Geographical studies of religion have almost universally proclaimed the Hispanic American borderland as a bastion of Catholicism (Shortridge 1976). Terry Jordan's mapping of church membership for Texas reveals the Texas-Mexican rimland to be a zone of Roman Catholic absolute majority (Jordan, Bean, and Holmes 1984, fig. 6.3). Other studies in San Antonio and South Texas confirm the strong Tejano attachment to Catholicism over many generations (Markides and Cole 1984; Juárez 1973). Less well-known is the im-

TABLE 7.1. *Selected Tejano Surnames in San Antonio and Los Angeles*

Name	San Antonio	Los Angeles
Arredondo	188	4
Benavides	204	35
Cadena	128	35
Cantú	525	19
Cavazos	178	6
Chapa	175	6
Elizondo	223	26
Garza	1667	75
Guajardo	173	10
Hinojosa	153	30
Medina	453	2
Móntez	102	18
Sáenz	238	64
Treviño	842	25
Villareal	740	44

SOURCES: San Antonio Residence White Pages, *Greater San Antonio* (Southwestern Bell 1994); Pacific Bell White Pages, *Greater Los Angeles* (Pacific Bell 1991). Greater Los Angeles directory includes only area code 213, which encompasses the central city and parts of East Los Angeles.

portant role of Protestantism in Tejano communities of South Texas (Remy 1970), and the religious tradition of charismatic folk-healers in the region (Romano V. 1965). Even more obscure, however, is the influence of Judaism in the Mexicano populations of northeast Mexico and South Texas (Larralde 1978). There is much evidence that Sephardic Jews were an important element of the colonial settlement of Monterrey and surrounding towns (Liebman 1970). This Jewish population assimilated after several generations, but certain cultural curiosities persist among Tejanos, whose roots were chiefly in northeast Mexico. Perhaps the most revealing relict of this past is the pastry *pan de semita*. This Lenten sweet bread is baked from wheat flour, water, butter, and mineral or vegetable oil. In Texas, pecans and raisins are sometimes included. That pan de semita (bread of Semites) is a carryover from early Sephardic Jews in the region is not well researched, but it may be one of a handful of culturally distinctive foodways (including *nopalitos lampreados*, nopal cactus and eggs, and *capirotada,* a sweet pastry) connected to a crypto-Jewish legacy of Hispanic South Texas (Montano 1992).

South Texas

Penetrated and explored as early as the sixteenth century, and crisscrossed by dozens of travelers during the nineteenth century, South Texas is, today, an acknowledged subarea of the Lone Star State. Texans are said to maintain a "perceptual image" of South Texas, especially as a directional region, notes geographer Terry G. Jordan, and historian Arnoldo De León labeled the region the "Tejano cultural zone" (Jordan 1978; De León 1982). The association of South Texas with its dominant Mexican American population—what geographers term a culture area—appears, however, to be a recent identification. In the nineteenth century, outsiders principally perceived the region as a wilderness rather than a settled area, and this image may have delayed its identity as a distinctive subcultural area until surprisingly late in this century (Arreola 1993e).

South Texas is the horn-shaped southern tip of the state. Its rough outline includes the lands south of the Balcones Escarpment, north and east of the Rio Grande, and west of the barrier-island coastline along the Gulf of Mexico. The northeast boundary is the San Antonio River. Colonial Spanish authorities first knew this region as Nuevo Santander. It included the current Mexican state of Tamaulipas and part of Coahuila, as well as present Texas south and west of the Nueces River. When Texas declared its independence in 1836, lands between the Nueces and the Rio Grande were disputed. While Texas claimed these lands, Mexico held that the territory had been part of Nuevo Santander, not Spanish or Mexican Texas that Texans had fought to proclaim their own. When Texas was annexed in 1845, this trans-Nueces disputed area became U.S. political territory, and by the 1880s South Texas was invaded by Anglo-Americans and their cultural influences (Arreola, 1992; Gonzalez 1930).

The horn-shaped area is unlike any other Hispanic American borderland environment. It is an east coast, subtropical semiarid zone that also has a riverine ecosystem along the Rio Grande. The lower portion of this area, commonly called "The Valley," is in fact an inland delta or embayment. Elevations immediately north of the lower Rio Grande actually decline to a point of inland drainage near the playa known as the Sal del Rey outside of Raymondville.

This coastal situation at a low latitude (circa 26°N) explains why the region's air is humid and sticky much of the year, and during the summers the humidity is even more uncomfortable because of high temperatures. Furthermore, there is very little physical relief along this coastal plain except for cuestas or escarpments that break the surface along northeast to southwest transects; elevations rarely exceed 500 feet except for northern parts of the area below the Balcones. The plant cover is a combination of grass and brush, technically part of the Tamaulipan biotic province (Blair 1950), but known locally as *monte*, *chaparral*, or *brasada*. The few streams that cross the horn in its northern reaches are exclusively spring sourced from the aquifer that underlies the Edwards Plateau above the Balcones Escarpment. Surface water is scarce between the Nueces and the Rio Grande.

The lower Valley of South Texas is some 250 miles south of San Antonio. As Robert Lee Maril (1992, 3) wrote, it is "as far south as you can get and still stand on the American mainland." Like the so-called Hispano Island of highland New Mexico (Nostrand 1992b), the horn of South Texas has been isolated from the rest of the United States, not by elevation but rather by sheer distance on this southernmost periphery of the continent.

These two conditions, environmental extremity and physical isolation, have combined to help shape the personality of la Tierra Tejana. Large-scale irrigated agriculture, a characteristic element of the modern identity of the region, is almost exclusively an Anglo-American enterprise. While some early Tejanos were farmers, the great shaping experience of the region has been the colonial and nineteenth-century livestock economy and way of life, known as *rancho* culture (O'Shea 1935; Jackson 1986; Graham 1994). As late as 1873, Tejanos still controlled some 157 ranchos along the American side of the Rio Grande from San Felipe (Del Rio) to below Brownsville (Martínez 1873). Folklorist Joe Graham (1991) has documented many ranching traditions that remain viable in the region including saddle-making, saddle-blanket weaving, bootmaking, blacksmithing, and several *vaquero* customs. In San Diego, a small town west of Corpus Christi, a vaquero foodways tradition—*pan de campo*—is the basis for a yearly fiesta (Arreola 1993d). Cattle culture is still important in the regional economy (Jordan 1993). When Anglo-American investors began to transform their ranching way of life, Tejanos, who had labored in the fields since the early twentieth century, embraced farming (Taylor 1934; Montejano 1987). Yet, despite the economic mainstay provided by migrant and field labor, this way of life has largely been a *peón* existence, and only with recent political rebellion and labor organization has the *bracero* become a heroic figure in the region (Foley et al. 1977). The romantic appeal of the *ranchero*, by comparison, has largely sustained Tejano identity, albeit transformed to an urban vaquero appeal through the enormous popularity of the musical sound early known as *conjunto* but today simply called *tejano* (Peña 1985; Adler and Padgett 1995).

In 1934 rural economist Paul Taylor characterized South Texas as an "American-Mexican Frontier," and thirty years later anthropologist William Madsen (1964) termed it a "Mexican-American" border subculture. Geographer Donald W. Meinig constructed in 1969 the first widely recognized, systematic geographic division of Texas by culture area rather than physical boundaries. Following sociologist Robert Talbert's work in 1955, Meinig mapped the stronghold nature of Mexican American South Texas based on counties with greater than 50 percent Spanish-surnamed population in 1960 and labeled the area as a "bi-cultural region."

More recent assessments continue to emphasize the Mexican and Tejano character of the region (Arreola 1993a, Roberts 1995). Furthermore, South Texas is culturally distinguished from other ethnic Mexican parts of Texas such as the trans-Pecos or West Texas and the high plains or Panhandle (Graham 1985; Haverluk 1993). Whereas the Mexican-origin population of the state is growing and expanding areally, especially into counties of the Panhandle, South Texas population growth has been less expansive spatially. Unlike the Hispano homeland (Nostrand 1992a), which expanded considerably during the nineteenth century, two Euro-Anglicized subregions areally restricted Tejano South Texas. To the north in the Edwards Plateau, the German hill country has been a non-Hispanic farming and ranching zone since the mid-nineteenth century, and to the northeast, Germans and Czechs in particular settled and cultivated the so-called Blackland string prairies (Jordan 1966, 1986). Given these non-Hispanic barriers along the northern and eastern edges of South Texas, and because South Texas borders Mexico, where the international boundary has remained relatively permeable, growth has not translated into territorial expansion. Rather, the open border leading to Texas seems to have created a safety-valve zone for Mexican immigrant employment opportunities. The Mexican migrant to the subregion brings cultural traditions from the fatherland but over the generations immigrant customs meld with the Tejano subculture. The result has been largely an intensification of the Tejano presence in long-settled nodes of the area, further entrenching and reinforcing the Tejano attachment to place.

Cultural Persistence across Three Centuries

Of the Hispanic American borderland subareas only Hispano New Mexico demonstrates greater antiquity and cultural persistence than does the Tejano homeland of South Texas. During the early-to-middle eighteenth century, Spaniards created colonial footholds in San Antonio, La Bahía (Goliad), and the Laredo area of the Rio Grande. In their usual town-founding tradition, they organized communities around plazas (Cruz 1988; Crouch, Garr, and Mundigo 1982). The starting point for towns like San Antonio and Laredo, for example, was the main plaza, and blocks, then streets, extended from this open space. Colonial plazas were the parade grounds and principal public gathering points.

During the nineteenth century, Anglo-Americans founded dozens of new communities in the region (Frantz and Cox 1988). Tejanos remained the majority population in many of these towns, and yet Anglos were responsible for laying out many South Texas Mexican-style plazas, although no single municipal code governed town founding or design as it had under the Laws of the Indies during the colonial era. At least five towns and perhaps others had land for the plaza donated by wealthy Anglo-American patrons (Arreola 1992). For example, John Twohig gave San Juan Plaza to the city of Eagle Pass when the town was laid out in 1850, and he stipulated that the land should not be used for any other purpose.

During the early twentieth century, South Texas, like other parts of the Hispanic American borderland, witnessed unprecedented immigration from Mexico (Gamio 1930; Cardoso 1980; Arreola 1993c). But South Texas was different from other borderland regions, in that Mexican movement to South Texas reinforced a Mexicano identity in towns. Because so many of these communities were already predominantly Mexican American, there was less resistance to such reinvigorated cultural ways than in other borderland subregions. Communities like Del Rio, Cotulla, and Hidalgo built plazas where none had existed previously, a gesture to the growing presence of Mexican-origin people in each town.

Plazas in South Texas towns functioned as traditional social nodes during the nineteenth and early twentieth centuries, especially as staging areas for the promenade, harvest and ranching fairs, and Mexican patriotic celebrations. Plaza towns are more numerous in South Texas than in the other Hispanic American borderland states: California, Arizona, and New Mexico. However, in New Mexico, Hispanos founded many agricultural villages as defensive plazas where houses were linked to create formal enclosures to protect against Indian raids (table 7.2).

TABLE 7.2. *Borderland Plaza Towns*

Town	Plaza Created
Santa Fe, NM	1610
Albuquerque, NM	1706
Laredo, TX	1767
Tucson, AZ*	1775
San Jose, CA	1777
Los Angeles, CA	1781
Santa Cruz, CA	1797
Socorro, NM*	1800
Monterey, CA*	1827
San Diego, CA*	1834
Las Vegas, NM	1835
Rio Grande City, TX	1846
Mesilla, NM	1848
Roma, TX	1848
Eagle Pass, TX	1850
Brownsville, TX	1862
San Ygnacio, TX	1874
San Diego, TX	1876
Benavides, TX	1880
Hebbronville, TX	1883
Alice, TX	1890
Zapata, TX	1898
Del Rio, TX	1908
Cotulla, TX	1925
Hidalgo, TX	1933

*Towns that evolved to include plazas.

SOURCES: Reps 1979; Crouch et al. 1982; Arreola 1992.

The long persistence of Mexicano cultural traditions in South Texas is evident as well from patterns of language use and community celebration. Nowhere in the Hispanic American borderland is the use of the Spanish language as resilient across such an extensive area as in South Texas, where some 19 counties are predominantly Spanish-speaking. Texas also counts more Spanish-language radio and television stations than its neighboring borderland states, including California, which has more Mexican Americans (Arreola 1995a). Finally, Mexican *fiestas patrias*, patriotic celebrations like Cinco de Mayo and Diez y Seis de Septiembre, as well as other Mexican religious and ethnic events are more common in Texas than elsewhere in the borderland. At least 47 Texas places host Mexican festivals today, with the greatest concentration in South Texas.

A Mexicano townscape, the persistence of spoken Spanish, and the popularity of ethnic Mexican celebrations give evidence of a long-standing South Texas region. Each of these manifestations has become even more pronounced as an ethnic revival has spread to towns throughout the area (Fishman 1985). Plazas are being restored and created anew and ethnic consciousness is perhaps at a high point.

Political Struggle and Control

Another identifiable quality that distinguishes the Tejano cultural heritage of South Texas is its political geography (Arreola 1993b). In 1994 Texas had the greatest number of Hispanic elected officials of any state (table 7.3). The bastion of this ethnic political might is chiefly in South Texas (map 7.3).

In a political sense South Texas is different from other areas in the state. Historically, the deep South Texas counties of Cameron, Hidalgo, Starr, Willacy, Kenedy, Kleberg, and Duval were the heart of the so-called *patrón* system, a semi-feudal arrangement derived from Hispanic colonial roots (Jordan, Bean, and Holmes 1984). The *patrón* was a political overlord who controlled ranch *peones* (peons) through social and economic patronage. In the early twentieth century this system survived almost exclusively in South Texas, where Anglo-American and Mexican American bosses like Jim Wells, Archie Parr, and Manuel Guerra built county-based political machines on the foundations of the older Hispanic ranching system (Anders 1982; Shelton 1974; McCleskey and Merrill 1973).

TABLE 7.3. *Borderland Hispanic Elected Officials, 1984–94*

Date	Texas	New Mexico	California	Arizona	Colorado
1984	1427	556	460	241	175
1985	1447	580	451	230	167
1986	1466	588	450	232	177
1987	1572	577	466	248	167
1988	1611	595	466	237	157
1989	1693	647	580	268	208
1990	1920	687	572	272	192
1991	1969	672	617	283	213
1992	1995	688	682	303	207
1993	2030	661	797	350	204
1994	2215	716	796	341	201

SOURCE: National Association of Latino Elected Officials, 1984–94.

South Texas has been a staging area for several Mexican American political movements. One was the short-lived Republic of the Rio Grande (1839–40), whose sovereignty encompassed the Mexican states of Tamaulipas, Coahuila, and Nuevo León, also the disputed territory between the Rio Grande and the Nueces River. Its capital was Laredo (Wilkinson 1975). A second was an irredentist movement sparked by the Plan of San Diego in 1915. Named after a small town in Duval County, this proclamation was likely drafted in Monterrey, Mexico. The manifesto called for a revolution against the United States to reclaim for Mexico land lost in 1836 and 1848—territory comprising Texas, New Mexico, Arizona, Colorado, and California. The rebellion was intended to be a race war, with every Anglo-American male over the age of 16 to be put to death. The leaders of the movement in San Antonio scheduled Texas to be liberated first and distributed handbills urging Tejanos to join them. For some ten months sporadic raids followed the proclamation causing havoc and forcing perhaps half of the population of the lower Rio Grande Valley to leave the region. Although the leaders of the uprising were to establish an interim republic across the Southwest with eventual reannexation to Mexico, the rebellion faltered before year's end (Sandos 1992).

The ethnic and racial overtones of early political struggles angered the rising middle-class Tejanos who saw themselves as separate from the Mexican American laboring class and migrants. As a result, the middle class began to

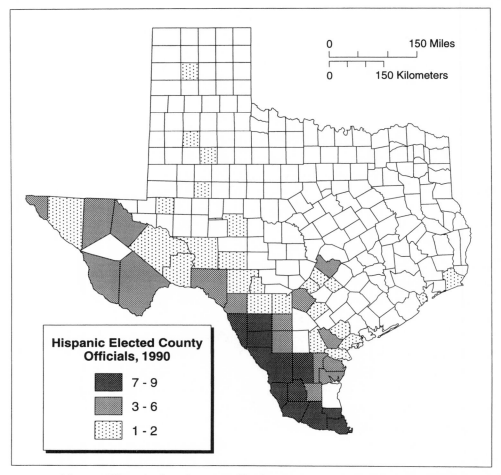

Map 7.3. Hispanic elected county officials (1990) are more numerous from South Texas than from other Hispanic subregions of the state. *Source:* National Association of Latino Elected and Appointed Officials, 1984–94.

identify itself increasingly as "Latin American" or "Americans of Latin American descent." This referent gained political legitimacy in 1929, when Tejanos formed the League of United Latin American Citizens (LULAC) in Corpus Christi (Márquez 1993). LULAC based its creed on two principles: Mexican consciousness in culture and social activity, but American consciousness in philosophy and politics. While LULAC favored the learning of the English language, it called for the maintenance of Spanish. Its agenda also advocated social and racial equality, the development of political power, and economic advancement. Although it was not a political party, it sponsored political, social, and cultural causes. LULAC currently has some 240 active councils and more than 200,000 members across the country, but it remains strongest in Texas (Kibbe 1946; García 1991; Meier and Rivera 1981).

During the 1960s South Texas gave birth to a Chicano movement. Some Tejanos separated themselves from the cultural pluralist posture of LULAC and inaugurated a radical political agenda. The catalyst for this awakening occurred when the control of local government in Zavala County shifted from Anglo-American to Tejano, an unprecedented event in the political history of the region (Shockley 1974; Foley et al. 1977). Crystal City, the self-proclaimed "Spinach Capital of the World," was then a small agricultural village in the Texas Winter Garden District. Some 85 percent of the town population was Tejano, yet the Anglo-American minority dominated all aspects of community life as it had since midwesterners founded the town in 1907. Politically powerless, the Tejano majority consisted chiefly of farm laborers, cannery workers, and seasonal migrants. In 1963 Teamsters Union members joined with Chicano activists to defeat the Anglo ruling elite in a local election that resulted in the establishment of Tejano political control in the community. Victory was short-lived, however; the Anglo minority regained control of government just two years later. In 1970 Tejanos organized La Raza Unida, a political party that proclaimed a platform of "Chicano nationalism" and gained control of the town council and school board. José Ángel Gutiérrez, a native of Crystal City and a cofounder of MAYO, the San Antonio–based Mexican American Youth Organization, steered the Chicano political victory (Miller 1975).

While the Raza Unida Party had some success in local and county elections in rural South Texas, it was unable to bring significant changes to this historically Democratic stronghold (McCleskey and Merrill 1973; Shelly, Archer, and Murauskas 1986). In 1972 a Raza Unida Party candidate, Ramsey Muñiz, of Corpus Christi, ran for governor but garnered only 6 percent of the vote and won only Brooks County. Representing the Raza Unida Party, Muñiz ran again in the 1974 gubernatorial election. This time he carried only Zavala County, although he finished second in 15 other Texas counties, 14 of which were in South Texas (García 1989; Jordan, Bean, and Holmes 1984).

These election results suggest the social and economic variability among Tejanos in the subregion despite their demographic dominance. For example, in border communities like Eagle Pass, Laredo, and Brownsville, places where Anglos and Tejanos have traditionally shared political leadership, Tejanos involved in local commerce have risen to middle- and even upper-class status. Because middle- and upper-class Tejanos lean to the two national political parties, especially the Democratic, the result has been that the Raza Unida Party fared poorly in these areas. As a consequence, Chicano political activity and the Raza Unida Party have had their greatest triumphs in inland towns such as Crystal City, Carrizo Springs, and Cotulla, where Anglos have traditionally controlled local economies and government (Miller 1975). Anglo minorities in these towns have long discriminated against Tejanos, and animosities served initially as the basis for ethnic politics, but, ultimately, third-party popularity was unable to carry the region. In 1978 the party captured less than 2 percent of votes in the gubernatorial election. Philosophical splits within the party and the formation of Mexican American Democrats (MAD) in 1976 led to the demise of the Raza Unida Party at the state level. Although efforts had been made to have the party become national, its heart and greatest strength were in Texas, especially South Texas. In 1982, when the party expired in its Crystal City birthplace, even its local influence came to an end.

A legacy of the Chicano political movement in South Texas is the Southwest Voter Registration and Education Project (SVREP), which Tejano Willie Velásquez, a cofounder of MAYO, formed and directed after he left the Raza Unida Party in 1974. Headquartered in San Antonio, SVREP has probably done more to empower Tejanos than any other single political initiative. Intended as a grass-roots organization to affect political change at the local level, SVREP brought together a coalition of civic, church, and neighborhood associations, labor groups, and volunteers to conduct door-to-door voter registration drives and education campaigns. In its first decade, SVREP helped increase the number of registered Hispanic voters in the Southwest by 1.6 million. In Texas alone the number increased from 488,000 to nearly one million (Brischetto 1988; Hufford 1988). Gains made in voter registration have not always directly translated to votes, however, because Tejano voter turnout has been low historically (DeSipio 1993).

One measure of the success of the SVREP in Texas is the 240 percent increase in the number of Tejanos elected to office between 1973 and 1990.

Nearly half of all publicly elected Hispanic officials in the United States in 1994 were in Texas (table 7.3). To a very great extent, the increase in the number of Tejano officeholders was the result of SVREP lawsuits that forced cities, counties, and school boards to change from at-large systems of election to single-member districts. Most of the Hispanic elected officials in the state hail from South Texas, including 745 school board members, 534 municipal officeholders, and 350 judges. All four U.S. Congressmen and 18 of 27 Texas state legislators who are Hispanic represent South Texas districts. Of Texas's 184 Hispanic county commissioners, clerks, treasurers, tax assessors, and other officials, 149, or 81 percent, hold office in South Texas (map 7.3). Only one South Texas core county, Kenedy, did not have an Hispanic county official. Kenedy is the least populated of the South Texas counties and it contains the headquarters of the famous King Ranch.

Place Bonding

Perhaps the key ingredient that defines a homeland is the bond that develops between a people and their place. In South Texas, that bond is evident in dozens of small towns, cities, and rural districts. Measuring the bond, however, is anything but precise. In large part, place attachment involves much that is unwritten because it is locked up in the hearts and minds of residents. Yet, because the landscape is, as Peirce Lewis (1979) called it, "our unwitting autobiography," then geographers should be able to decipher from its necessarily incomplete clues something about a sense of place. I have selected two places in South Texas to assess this bond between Tejanos and their *tierra*. Appropriately, each place is named for a Catholic saint: Antony of Padua and Ignatius Loyola. One vignette examines the place attachment to a commercial district of a Tejano city over a century, and the other assesses the close familial attachment to a South Texas village, one that illustrates a striking vernacular built environment for nineteenth-century Tejano culture.

El West Side

San Antonio is the "Mexican American Cultural Capital" (Arreola 1987). Settled in 1718, more than a century after Santa Fe in the Spanish borderlands, San Antonio emerged as a center of Mexican, not Spanish, culture in the

United States during the late nineteenth and early twentieth centuries, when its strongest links were forged with Mexico. Throughout much of its history, the Texas city has been the center of Mexican heritage and cultural innovation in the U.S. Mexican cultural dominance results from its homogeneous ethnic character—Mexican Americans are unchallenged by other Hispanic groups—and its percentage rank among cities with large Mexican American populations: 59 percent of the city was Mexican in origin in 2000. Whereas "Mexican American" is still used as a subgroup designation in San Antonio, increasingly, Tejano challenges this label (Institute of Texan Cultures 1971; Poyo and Hinojosa 1991; Matovina 1995).

The heart of Tejano San Antonio is the so-called El West Side. Although Mexican Americans are present in most parts of the city, including especially the central, southwest, and southern sectors, the West Side is the spiritual home of Tejanos in San Antonio (West 1981; Garrett 1988). The West Side formed during the early twentieth century, when immigration from Mexico flooded the city with new residents who concentrated in barrios called Laredo and Chihuahua west of San Pedro Creek. Crossing this creek west of Military Plaza (present City Hall) was the equivalent of fording the Rio Grande. In 1938 city guidebooks declared that San Pedro Creek, long a line of demarcation, was the point at which the Mexican quarter of San Antonio began. Beyond this line, an area of slightly more than a square mile contained almost three-fourths of the Mexican population in the city by 1940 (Arreola 1995b).

The commercial focus of the West Side, the heart of the quarter, was a ten-block district that encompassed several squares and the chief Tejano business community (map 7.4). The squares or plazas were important social gathering points for Tejano San Antonio, especially Haymarket Square, which served as the meeting point for wandering bands of Mexican minstrels and became the famous setting for outdoor chili stands until 1937. At Milam Square men seeking day labor would typically gather, and it became the focus of a civic event known as "Night in Mexico," where local organizations sponsored Mexican folk dances and other traditional celebrations.

East of the squares was a six-block area that contained the highest concentration of Tejano-operated and patronized businesses in the city. Because San Antonio was recognized in Mexico and across the southwestern states as a major commercial node for a vast hinterland and a collecting point for migrant labor, its Mexican business district ranked second to none (García 1991).

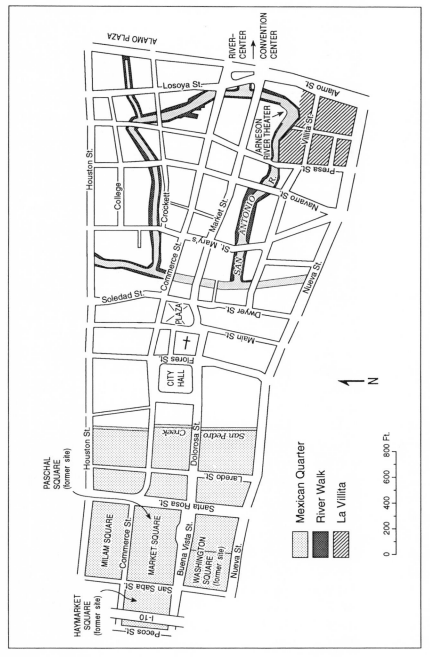

Map 7.4. San Antonio's Mexican commercial quarter (1995) consists of a ten-block district west of San Pedro Creek. *Source:* Arreola 1995b.

TABLE 7.4. *Tejano-Operated Businesses in San Antonio's Mexican Downtown, 1924*

barbería (barber shop)	14
restaurante (restaurant)	9
sastrería (tailor shop)	9
abarrote (grocery store)	7
calzado (shoe store)	6
joyería (jewelry store)	6
droguería (drug store)	5
hotel (hotel)	5
librería (book store)	3
música (musician's store)	2
imprenta (printing shop)	2
pandería (bakery)	1
carpintería (carpenter's shop)	1
segunda mano (second hand store)	1
quiropráctico (chiropractor)	1
club (social club)	1

NOTE: n = 73.
SOURCE: Sologaistoa 1924, 211–22.

A 1924 directory listed 73 Mexican businesses on these six blocks alone, with the greatest number of establishments along Laredo Street between Houston and Nueva streets (table 7.4). The Lozano family published the Spanish-language daily *La Prensa* (1913–42) in this district, and the newspaper boasted that it maintained the largest circulation of any "Mexican" newspaper in the United States (*La Prensa* 1923). Advertisements in this paper show that Tejano businessmen in San Antonio were supplied by importers from Eagle Pass and Laredo on the Rio Grande with connections to such northeast Mexican cities as Piedras Negras, Nuevo Laredo, Allende, Cuatro Ciénegas, Saltillo, Monclova, Múzquiz, and Sabinas. Monterrey, the large industrial node of northeast Mexico, had been an important commercial link since the completion of the railroad to that city in 1882 (Sánchez 1898).

The creation of this business focus on the west side of downtown San Antonio stamped the district in the consciousness of residents as the Mexican downtown. Urban renewal threatened the quarter in the 1960s and many of the businesses closed or relocated to the suburban West Side. In the 1970s a joint public and private investment effort by the City of San Antonio and the San Antonio Development Agency spawned the conversion of streets into pedestrian malls with fountains, landscaping, and ornamental lights and benches. El Mercado, as it is officially known, encompasses a large enclosed Mexican crafts market, tourist boutiques, and eateries, including the very popular restaurant-and-bakery, Mi Tierra. On any given day the square attracts locals as well as out-of-towners and becomes especially crowded during festive celebrations when artists and the latter-day chili queens set up their stands to tempt passersby. In some ways, El Mercado is the modern version of the Military Plaza that hosted nineteenth-century market and social activities, combined with the functions of wandering musicians and "Night in Mexico" celebrations that Tejanos staged on Haymarket and Milam squares during the 1920s to 1940s (Arreola 1995b). This attachment to place has persisted for over a century, signaling a bond between Tejanos and an urban social space.

Stone and Spirit

Before European brick making became popular in South Texas, sandstone and limestone cut from shallow quarries were popular building materials. Caliche in this region, cut typically into large blocks, is termed *sillar*. In construction it is usually chinked with smaller rocks, mud, and cement to fill spaces between the layers of stone, a process sometimes called *rejoneado* or *ripio* (Newton 1964; George 1975). The village of San Ygnacio, some 35 miles south of Laredo on the Rio Grande, contains some of the finest surviving examples of sillar-construction in all of South Texas (Robinson 1979; Graham 1992). In San Ygnacio, this vernacular built environment combines with a familial legacy that gives the small town a special identity in the region, an identity that captures the roots of Tejano culture in northeast Mexico and preserves this patrimony in stone and spirit.

San Ygnacio, the oldest town in Zapata County, was founded as a subdivision of the José Vásquez de Borrego land grant in 1830. Under the leadership of Don Jesús Treviño, residents from Revilla, Tamaulipas, first settled the town (Fish 1990). (San Ygnacio was the patron saint of the town of Revilla, now old Guerrero, which lies under Lake Falcon; see Byfield 1966 and McVey 1988.) Treviño selected a site on a sandy level plain south of the Arroyo Grullo along the banks of the Rio Grande, where he constructed a stone house. Added to over the years, it stands as the oldest structure in town.

The early population of San Ygnacio was almost exclusively composed of transplanted residents from Revilla, and within a single generation the Treviño and Uribe families intermarried. Don Blas María Uribe became the town patriarch and a successful businessman, establishing a train of pack mules that hauled goods between Corpus Christi and San Ygnacio as well as a line of freight boats that navigated the Rio Grande seasonally between San Ygnacio and Brownsville (Fish 1990; Kelly 1986). Before the railroad era in the late nineteenth century, San Ygnacio functioned as a pivot in trade between South Texas and Mexican markets in Monterrey, Monclova, and Saltillo. Texas cattle moved west and south into Mexico while beans, flour, corn, *piloncillo* (raw sugar), and other staples were traded east and north across the river.

Until the 1870s, San Ygnacio remained a small settlement of a few sandstone buildings. In 1872 Uribe donated land for a church, Nuestra Señora del Refugio, and a town plaza. The town was finally platted in 1874 (Barbee 1981). The residents arranged the community, labeled "Rancho de San Ignacio" on the plat, in a grid of some 20 blocks centered on the plaza, known today as Plaza del Pueblo (Arreola 1992) (fig. 7.1). By 1917 the town had a population of 500 residents, and by 1951 it had increased to 1,000; the community is approximately the same size today as it was in the 1950s (Pierce 1917; Lott and Martínez 1953).

San Ygnacio remains a villagelike place, proud of its Tejano heritage, which it nurtures in several ways. On every Good Friday since 1851, the town's residents have celebrated the procession of the "Via Dolorosa" symbolizing Christ's crucifixion walk. To commemorate this event, 14 stations of the cross are set up at specific points linked to San Ygnacio's prominent families. Each family is responsible for adorning a small table at the station stop in the middle of the street. The procession begins at the Nuestra Señora del Refugio church on the plaza and winds through the town. A life-sized wooden cross is carried at the front of the parade, followed by four young girls who carry the statue of Nuestra Señora del Refugio on a litter, then a priest and town residents. The procession pauses at each station table where the celebrants place the statue of the virgin and say prayers. As the parade moves through the town, participants sing between station stops. The event is completed when the procession returns to the church and a Mass is said.

Fig. 7.2. Adrian Martínez house (left, restored) built in 1873 by Manuel María Uribe, and side view (right) of Jesús Treviño house built in 1830. The San Ygnacio Historic District includes 36 sandstone structures built in the nineteenth century. Photograph by Daniel D. Arreola, 1993.

Since at least 1964, when the Treviño house, built in 1830 by the town founder and added to by Uribe, was designated a Recorded Texas Historic Landmark, the stone buildings of San Ygnacio have attracted increasing regional, national, and international attention (fig. 7.2). In 1973, 36 stone buildings scattered throughout the town became a part of the San Ygnacio Historic District and appeared in the listing of National Register of Historic Places (Sánchez 1991). According to architectural historian William Barbee (1981), workers quarried sandstone, known as *piedra de arena*, east of town from hillsides where it is near the surface. Masons assembled the building pieces with a mortar mud called *zoquete*. A plaster, *enjarre*, made from white sand from the Arroyo Grullo and lime-kilned locally, sealed the stone.

San Ygnacio continues to take pride in its vernacular built environment and Tejano cultural heritage. In 1982 the Zapata County Historical Commission formally dedicated the La Paz Museum in San Ygnacio as the official county repository (Fish 1990). In 1987 San Ygnacio hosted a historic celebration in which dignitaries and residents participated in a weekend commemoration of the town's 157 years, honoring both historic and contemporary families.

Conclusion

The South Texas Tejano homeland, like the Hispano homeland of New Mexico, and the southern Arizona homeland focused on Tucson, is one of several Hispanic American borderland subregions. Each of these homelands has roots in Mexico, or New Spain, yet their persistence in the borderland and the geographic isolation that long separated them have given rise to subregional identities with distinctive cultural characteristics.

In la Tierra Tejana, an area larger than the state of Pennsylvania, Tejanos have shaped a special cultural identity. For nearly three centuries, Hispanic influences have spread over and imprinted this landscape of dry plains and low hills. From Spanish place names to a ranching way of life, from Mexicano townscapes to ethnic politics, Tejanos have bonded with and branded this land. Their influence is evident even in the Anglo-Texan communities of the region, where Tejano foodways, language, ranching lifestyle, and ethnic politics have spiced land and life. Sheer Tejano demographic dominance ensures the perpetuation of this homeland. Its proximity to the Mexican source area across the international boundary and the interaction across that boundary of people, goods, and cultural ways will sustain the Mexicano character of la Tierra Tejana.

CHAPTER EIGHT

The Anglo-Texan Homeland

. . . .

Terry G. Jordan-Bychkov

A *homeland*, to me, is a region long inhabited by a self-conscious group exercising some measure of social, economic, and political control over the territory while at the same time not enjoying or even seeking full independence. The group exhibits a strong sense of attachment to the region and has created special, venerated places that symbolize and celebrate their identity. Usually peripheral in location, homelands combine the attributes of formal and functional culture regions, becoming in the process potent geographical entities. They are incompletely developed nation-states. Ethnic status is not a prerequisite for homeland formation, as exemplified by Bavaria, Andalucía, Tuscany, Scotland, and any number of other regions.

Anglo-Texans, I suggest, fit this model of a homeland (Jordan 1993b). They constitute a self-conscious, proud group strongly attached to place, occupy a peripheral location within a larger country, possess profane shrines dedicated to their identity with a corollary historical mythology, and exercise considerable social, economic, and political control over their territory without having or wanting independence. We may define an Anglo-Texan as a person of old-stock American origin having at least partial British ancestry. The term "Anglo" is employed in Texas more broadly to mean any person not of Hispanic, African, or Asian blood, but the Anglo-Texan homeland is rooted in the more restricted definition of "WASP," or white Anglo-Saxon Protestants derived from the American South.

Anecdotal and Other Evidence

Evidence of the homeland status of Anglo-Texan culture, while abundant, is often anecdotal. One repeatedly hears, for example, the story of Texans forced by circumstance to live outside the state, who, at the birth of a child, import a box of Texas soil for the doctor to stand in while delivering the baby. Similarly, the other great life event—death—often prompts nonresident Texans to ask for burial in the state. One of my uncles, who spent his adult life in Georgia, requested when near death to be interred in the East Texas graveyard near his boyhood home, a thousand miles from his grieving, non-Texan family.

Bumper stickers—one of the more revealing if largely unstudied messages of popular culture—also often speak of Anglo-Texan identity. "Secede!" plead many, half-jokingly; "Native Texan" boast others. "Let them freeze in the dark," a favored 1970s message, was as sincere and quintessentially Anglo-Texan as "Remember the Alamo." "Yankee go home" appeared during the oil boom–induced wave of immigration from the north in the early 1980s, a time when some number-crunching Texan economist reckoned that, if independent, Texas would have a GNP ranking in the top ten worldwide. A restaurant in Belton boasts that it is "Texan Owned," and Fords assembled at a Dallas auto plant in the 1940s and 1950s bore the window label "Made in Texas by Texans." Repeatedly the keepers of the homeland faith remind Texans that they possess (often apocryphal) rights not enjoyed by other states: that the Lone Star banner can be flown at the same height as the American flag; that the state rather than the federal government owns and draws oil wealth from the lands of the public domain, including offshore "tidelands"; or that Texas has the right unilaterally to divide into as many as five states.

At the same time, Anglo-Texans desire neither independence nor fragmentation. Most understand that, contrary to the stereotyped image, Texas is a rather poor state. Only once, at the brief peak of the oil boom, did the state's income per family reach the national average, and Texans receive far more from the federal government in Washington than they pay in federal taxes. Indeed, the state has been treated virtually as a depressed region requiring special federal assistance. In return, Anglo-Texans became American super-patriots, always ready to die in whatever jungle or desert Washington posts them to.

Seventh-grade students in the Texas public schools must study the state's history for a semester, an experience that, astoundingly, state law encourages be repeated at every tax-supported institution of higher learning. Until recently these courses entailed a blatant indoctrination in Anglo-Texan mythology. I recall seeing, as a public school child in the 1940s, a particularly chilling documentary about the Alamo battle, in which, on the fateful final day of the siege, the armed alien horde suddenly appeared at all points along the horizon. The camera slowly panned 360 degrees to reveal an unbroken line of what must have been a million swarthy Mexican soldiers, standing menacingly with fixed bayonets in a dead calm before the final assault. Though today saccharine, politically correct drivel has replaced such hate propaganda, the old lessons were so well taught and learned that the mythology remains alive and well. An imprudent revisionist historian not long ago wrote a book pretty much proving that Davy Crockett had surrendered at the Alamo rather than going down fighting (Kilgore 1978). The historian, predictably, became the target of insults, hate mail, and midnight calls. As further expressions of the penchant for self-study, Texans have their own state almanac and a remarkable six-volume state historical encyclopedia. Texana is much sought by the state's book, art, document, and antique collectors, so much that counterfeiting and thievery thrive.

Texas toponyms reinforce Anglo mythology. Streets, towns, and counties bear the names of the founding fathers (sorry, no mothers), heroes, and martyrs of Anglo-Texan colonization and independence from Mexico, toponyms such as Austin, Houston, Travis, Crockett, Bowie, Fannin, and Deaf Smith (map 8.1). Ninety-one counties bear such names, while Texans named only 13 for prominent Americans and but five for Confederates. Complementing these toponyms are an assortment of homeland shrines (map 8.2). Few Anglo-Texans have not visited the Alamo (their "Thermopylae," as teachers used to say before classical knowledge vanished) and the San Jacinto battlefield with its enormous victory pillar (early efforts to Anglicize the name to St. Hyacinth failed, and Anglos must content themselves with grossly mispronouncing it). Washington-on-the-Brazos, where Texans signed their declaration of independence in 1836, provides another shrine, as do Sam Houston's home, grave, and colossal 67-foot tall statue in Huntsville (fig. 8.1) and the magnificent state capitol building in Austin, its dome towering higher than the one in Washington, D.C. As a boy, I was taken to most of these places, a pilgrimage experience not unusual for middle-class Anglo-Texans.

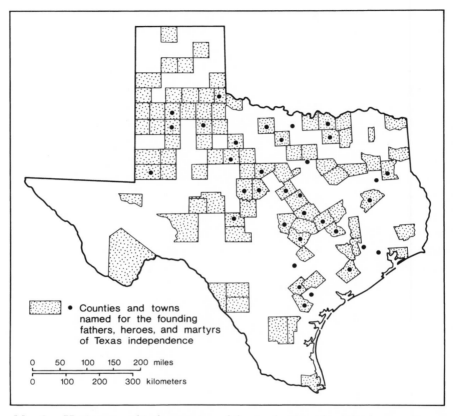

Map 8.1. Hagiogeographical toponyms of the Anglo-Texan homeland, 1952. *Source:* Webb and Carroll 1952.

Origins of a Homeland Mentality

Why are Anglo-Texans not like other WASPs? Why did they develop a self-conscious identity and homeland mentality? After all, they trace their genealogies to relatively unremarkable states such as Tennessee, which sent not just its sons and daughters, but also the hill twang Anglo-Texans like to regard as their own unique dialect, and Alabama, as much a part of the Cotton and Bible belts as Anglo-Texas. More remotely, Texan genealogies reach back to Virginia, Pennsylvania, and South Carolina, the states that mothered the South at large.

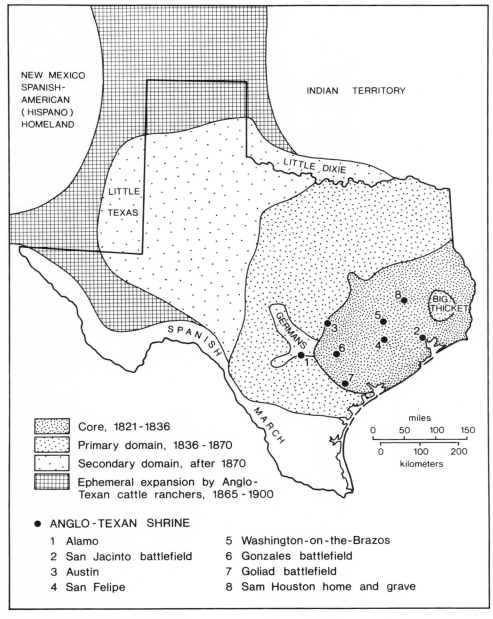

NEW MEXICO
SPANISH-
AMERICAN
(HISPANO)
HOMELAND

INDIAN   TERRITORY

LITTLE DIXIE

LITTLE

TEXAS

GERMANS

BIG
THICKET

SPANISH

MARCH

8

5

3

2

6

4

1

7

miles

0   50   100   150

0   100   200

kilometers

Core, 1821-1836

Primary domain, 1836-1870

Secondary domain, after 1870

Ephemeral expansion by Anglo-
Texan cattle ranchers, 1865-1900

● ANGLO-TEXAN SHRINE

1 Alamo   5 Washington-on-the-Brazos
2 San Jacinto battlefield   6 Gonzales battlefield
3 Austin   7 Goliad battlefield
4 San Felipe   8 Sam Houston home and grave

Map 8.2. The Anglo-Texan homeland, nineteenth century. Origin, expansion, and secular shrines.

Fig. 8.1. Statue of Texas hero Sam Houston captures symbolically the Anglo-Texan homeland. This 67-foot-tall statue—reputedly the largest freestanding sculpture in North America—towers over Interstate Highway 45 near Huntsville, where Houston's home and grave provide one of the secular shrines of Anglo-Texas. Photograph by Terry G. Jordan-Bychkov, 1996.

Anglo-Texan identity and homeland spring from several causes. Most crucially important were violent encounters with alien cultures—Hispanic and, to a lesser extent, Comanche. Texans were the first Anglos to experience major and prolonged unpleasant contacts with long-established Latin Americans and mounted Plains Indians. Nothing raises self-consciousness as surely as the clash with very different peoples. As a result, Anglo-Texans became aware, to an extent never experienced by other southerners, who they were and were not. The Anglos eventually destroyed the Comanches and seized their lands, though the struggle required forty years and there were times when the outcome was in doubt, as in 1840, when the Comanches drove all the way to the Texas coast (Fehrenbach 1974). Remarkably, at the height of that bloody struggle the Anglos named a county for the Comanches, an act roughly equivalent to naming one for Hirohito in 1942.

The encounter with Mexican culture, while resolved militarily much earlier than the Indian wars, proved far more important in shaping Anglo-Texan self-identity and chauvinism. In the final analysis, the Anglos took almost no territory from the Mexicans in the cultural sense: the border between Anglo and Latin America today lies where it did in 1836, along the axis of the San Antonio River, the border of Daniel Arreola's Tierra Tejana. If Anglos took little territory culturally, they acquired even less from Mexican culture. True, Anglos happily consume a bland, acculturated version of Mexican food and habitually utter a few mispronounced Spanish loanwords, usually ignorant of their origin. But meaningful cultural exchange and blending have not occurred. Intermarriage remains uncommon and Anglo rejection of borderland Tejano music is virtually complete, even though this sound reflects a mixture of Mexican, German, Polish, and Czech influences and probably evolved in south-central Texas (Peña 1985). Similarly, Mexican folk Catholicism, rich in pilgrim shrines, visions of the Virgin of Guadalupe, and associated statuary, is generally viewed with scorn and amusement by Anglo Bible-Belters. The great majority of Anglos clings to a defiant, xenophobic monolingualism, so profound as to render meaningless such venerable Texas place names as Nueces, Lampasas, Sabine, Llano, and Pedernales. Bilingualism is regarded by most Anglos as undesirable and threatening. Indeed, threat is precisely the role played, however unintentionally, by Mexicans for more than a century and a half. They represent a brown peril at the gates of this peripheral outpost of Anglo-Saxon culture, engendering a sense of arrogant pride, self-consciousness, and cultural siege. Music, religion, and mother tongue all threaten. Latin America created the Anglo-Texan homeland.

Anglo-Texan identity was further reinforced by the loss of the Civil War. Though most Texans became enthusiastic Confederates, contributing mightily to the southern military effort, the loss of the war did not set well in the Lone Star state. Perhaps, at least subconsciously, the specter of the Alamo was raised by Appomattox. While the likes of Alabamians and Virginians settled into nostalgia for the Lost Cause, content to hate Yankees and gently decay, Texans chose instead, over the course of several generations, to renounce the South and embrace the West, thereby severing themselves from their cultural roots and providing a convenient eastern border for their homeland. Few Anglo-Texans today regard themselves as southerners (Reed 1976, 932; Zelinsky 1980, 14). The quasi-French character of much of Louisiana reinforces the homeland's eastern boundary, and some measure of truth resides in the Cajun wisecrack that "the Sabine divides the coonasses from the assholes."

Shifting demographic patterns have recently renewed and heightened the old fear of Mexicans that keeps the Anglo-Texan homeland viable. In 1887 Spanish-surnamed persons formed only 4 percent of the Texas population, but the proportion rose to 12 percent in 1930, 15 percent by 1960, and 26 percent in 1990, mainly as a result of substantial, protracted immigration (Foster 1889; Jordan 1986). If *Anglo-Texan* defines a person of at least partial British ancestry, then this group had dwindled from close to a two-thirds majority in the state in the late nineteenth century to a mere plurality as early as 1980. This transition will become increasingly traumatic for Anglos as the new demographic order in Texas continues to be translated into political terms. Even the ongoing assimilation of former ethnic continental Europeans and Yankee immigrants into the Anglo population can scarcely bring back the old days. As the state slips from their grip, the homeland may become increasingly important to them.

Homeland Development

The political boundaries of the state of Texas do not coincide with the Anglo-Texan homeland, nor have they ever. Anglos have traditionally used the functional apparatus of the state to control and, for a time, enlarge the homeland, but the two have never been coextensive. Even so, every governor since statehood in 1845 has been Anglo-Texan, suggesting the functional importance of the state to the homeland. Donald Meinig viewed this arrangement as quasi-imperial, and for a time it was (1969).

The nucleus of the Anglo-Texans' homeland lay in southeast Texas where their cultural identity was forged in the formative period 1821–36, in contact with and under political rule by Mexicans. The essential we-they mindset that ever after defined Anglo-Texan self-consciousness took shape then and there, rather than in the decade of reluctant political independence that followed. Of the Anglo-Texan shrines, only one—the Alamo—lay outside this core, situated in a city that, then as now, remained just beyond the full grasp of the Anglos. If I were forced to draw a sharp southern border for the Anglo homeland (a fool's errand if ever there was one) the line would have to pass through San Antonio, the Alamo Plaza, and the old mission/fort itself. Almost central in this core area lay San Felipe, the capital of Stephen F. Austin's colony, the main focus of Anglo settlement during the Mexican period. San Felipe, though totally destroyed during the war in 1836 and never rebuilt, retained enough residual symbolic meaning to Anglos that citizen outrage greeted the state's razing of the replica of Austin's log cabin there—a hokey, latter-day structure built of surplus telephone poles.

Following the war of independence, Anglo-Texans steadily expanded their homeland to the north and west, expelling the Cherokees from northeast Texas and gradually annexing the Comanchería. Eventually they spilled over the political boundaries of the state into areas such as "Little Texas" in southeastern New Mexico and "Little Dixie" in southern Oklahoma. With the support of the United States Army, they created a latter-day Spanish March in the south, one millennium after Charlemagne. That buffer zone would remain culturally part of Latin America but usually militarily secure, garrisoned by the army and patrolled by the hated Texas Rangers. To the north lay the Indian Territory of Oklahoma's Five Civilized Nations, while in the west the Anglo-Texans eventually bordered the Hispano or Spanish-American homeland in highland New Mexico (Nostrand 1980, 1992; Meinig 1971, 74–91; Carlson 1990).

The impressive expansion of the Anglo-Texan homeland did not continue beyond the turn of the century. Advance soon gave way to retreat, retrenchment, and finally to a clear sense of cultural siege. The retreat began in the year 1896, when the United States Supreme Court awarded Greer County to

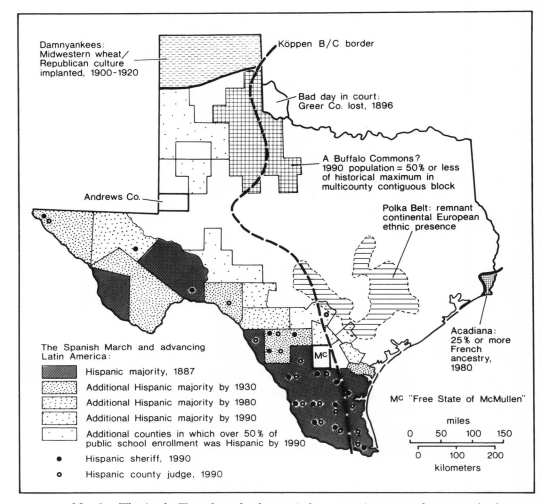

Map 8.3. The Anglo-Texan homeland, twentieth century. A century of retreat and cultural siege. *Sources:* Foster 1889; Kingston 1991.

Oklahoma Territory, ending a long dispute over which fork of the Red River served as the northern border of Texas (map 8.3). Created by the Texas legislature in 1860, organized in 1886, and peopled largely by Anglo-Texans, Greer County almost certainly would have remained with Texas, had a plebiscite been allowed. "Greer County, Texas" is repeatedly carved on the granite markers that today line the bizarre pioneer Hall of Fame near the courthouse in the county seat, Mangum (fig. 8.2). Though old Greer County would later give Anglo-Texans one of their popular heroes—football coach Darrell Royal—it was lost forever to the homeland.

Similarly, the creation of state governments in Oklahoma and New Mexico in 1907 and 1912 sealed the fate of Little Dixie and Little Texas. Falling beyond the pale of the functional political region, they, as surely as Greer County, were lost. No citizen of Texas today thinks of these territories as being in any sense Texan. Likewise, the northward migration of cattle ranchers in the Great Plains after 1865, creating a "Texas Extended" as far as the Cypress Hills of Alberta, formed a diaspora the homeland never claimed (Jordan 1993a, 208–40).

In fact, even that portion of the Great Plains lying within the political boundaries of Texas could not be retained in its entirety. After 1900 in the northern Panhandle, waves of midwesterners bearing a wheat-Republican culture overwhelmed and displaced Anglo-Texan cattle ranchers. Not even the subsequent discovery of oil there, with a resultant intrusion of Anglo-Texan roughnecks in towns such as Pampa, could effectively reclaim the northern Panhandle for the homeland (Meinig 1969, 106–7). Recently the 26-county Panhandle only half-jokingly moved toward secession from Texas. In far southeast Texas, that same oil era lured in a Louisiana Cajun workforce, making counties such as Orange and Jefferson culturally almost as much a part of Acadiana as Anglo-Texas.

The most substantial and traumatic retreat has been the result of the mass immigration of Latin Americans, mainly Mexicans, since about 1900, with the resultant northward advance of Hispanic culture and Anglo loss of political control over the Spanish March in the south (map 8.3). With each passing decade, more counties acquired Hispanic majorities and pluralities. The magnitude of the majority has grown, further aided by Anglo emigration, so that some counties of the marchland are now over 90 percent Hispanic. A political awakening, triggered by events at Crystal City in South Texas in the 1960s, allowed the Hispanic element to wrest local governmental control of the marchland from the Anglos (Shockley, 1974). Most counties with Hispanic majorities now also have Spanish-surnamed sheriffs and county judges—the most sensitive positions. No longer do landed Anglo *patrones* and the Texas Rangers rule the borderland.

The northward advance of Hispanic majorities and political power has

Fig. 8.2. Pioneer Hall of Fame, Mangum, Oklahoma, in old Greer County, "Texas."
Photograph by Terry G. Jordan-Bychkov, 1990.

Fig. 8.3. Sign announcing the "Free State" of McMullen County, Texas, bears a hidden meaning and reveals an outpost mentality. Photograph by Terry G. Jordan-Bychkov, 1987.

caused a siege mentality to develop among Anglos, especially near the cultural "front." Residents of McMullen County in South Texas, now almost completely surrounded by Hispanic majority counties, half-jokingly call it the "Free State of McMullen" (fig. 8.3). Though local Anglos deny that this appellation bears any ethnic message, the Free State in fact reflects a Nagorno-Karabakh mentality. Its real meaning is the "Anglo State of McMullen." The Free State's outpost days are apparently numbered, for while the county's population was only 39 percent Hispanic in 1990, the public school enrollment revealed a Hispanic majority for the first time in 1991 (Kingston 1991, 182, 269; Garcia 1991).

The most spectacular Hispanic advance has been in the South Plains or Llano Estacado of West Texas. There, Mexican Americans have recently achieved majorities in public school enrollment in 12 counties, including Deaf Smith, named for a hero of San Jacinto. The Anglo-Texan homeland, it seems, is destined to lose all its semiarid areas, and the cultural border may eventually stabilize along the climate boundary. That is probably a fitting divide, for Anglos have usually behaved maladaptively in moisture-deficient areas.

This emerging new homeland border within the state is well buffered. In the south-central part of Texas, incompletely assimilated continental Europeans still dominate most of the seam between Anglos and Mexicans. While the long-term destiny of these Polka-Belt minorities seems almost certainly linked to the Anglo-Texan community, a special cultural identity will persist there for at least another generation. In the northwest, Anglos have for a half-century been withdrawing from a sizable block of hardscrabble counties below the Caprock escarpment, producing de facto a Popperian "buffalo commons" on the margins of the increasingly Hispanicized High Plains. Four of the hardscrabble counties by 1990 had less than one-fourth of their historical population maxima, and all housed under one-half of that total.

In the much reduced Anglo-Texan homeland, retreat and cultural siege have somewhat revived cultural identity. Contact and conflict with Hispanics today, as in 1821–36, reinforce Anglo self-consciousness. In common with Serbs, Armenians, and Catalonians, Anglo-Texans still know who they are.

CHAPTER NINE

The Kiowa Homeland in Oklahoma

· · · · · · · · · · · · · · · · · · · ·

Steven M. Schnell

Homeland Background

The earliest place where the Kiowas are known to have lived is the northern Rocky Mountains, near the headwaters of the Yellowstone and Missouri rivers. Unlike Cheyennes and Arapahoes, Kiowas have no memory of ever having been an agricultural people. Sometime prior to 1700, the tribe moved east out of the mountains into the Black Hills area. About this same time, they acquired horses, probably from the neighboring Crows, and began to develop the buffalo-hunting culture that was to define them as a people for future generations (Mooney 1979) (fig. 9.1).

In about 1750, Cheyennes and Lakotas forced the Kiowas from the Black Hills, and they began a long, gradual southward migration. By 1833, Kiowas had centered their lives near the Wichita Mountains in southwestern Oklahoma, and along with their allies the Comanches, they soon held firm control over the southern Plains from the Arkansas River into central Texas, and from the Cross Timbers of central Oklahoma west into the Llano Estacado of the Texas Panhandle. In 1867, the Medicine Lodge Treaty required Kiowas to settle on a reservation (along with the Comanches and Plains Apaches, who are usually misleadingly referred to as Kiowa-Apaches) in southwestern Oklahoma. Their new reservation was a fraction of the size of their previous range (Mooney 1979) (map 9.1).

Fig. 9.1. The way to Rainy Mountain. Photograph by Steven M. Schnell, August 1999.

Even this reduced range was not to remain theirs for long. As the so-called Boomers agitated for opening lands in Indian Territory to white settlement, Congress passed the Dawes Severalty Act, which provided for the dissolution of tribal reservations. The legislation allotted each Indian a 160-acre homestead; the government purchased the remainder of the land and opened it to white settlement in 1901. While Kiowas settled almost exclusively along creeks north of the Wichita Mountains, Comanches tended to take their allotments south of the modern town of Apache and farther south of the mountains. Members of the much smaller Plains Apache tribe generally chose land along a strip running roughly from Apache north to the intersection of today's Oklahoma Route 9 and US Route 62/281 just south of Washita. By the time of the last allotment, the amount of land in Indian hands had shrunk by two-thirds, to 443,338 acres (Mayhall 1971, 319). Land sales and outright swindling soon deprived tribal members of many more acres.

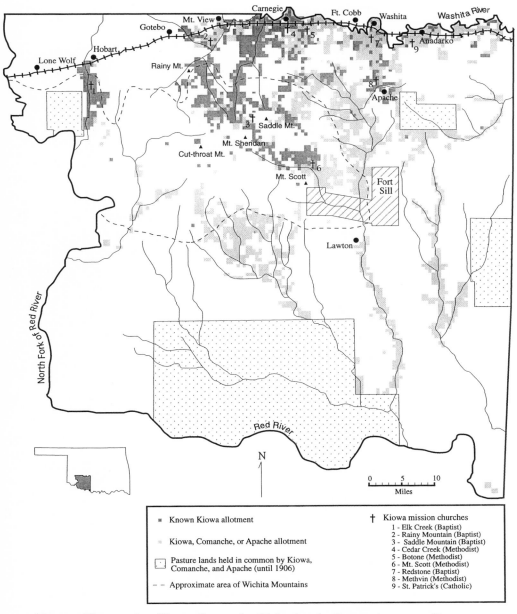

Map 9.1. The one-time Kiowa–Comanche–Plains Apache Reservation, 1900. Shown are Kiowa (also Comanche and Plains Apache) allotments made in 1901 and various landmarks including mission churches. *Sources:* Griffin 1901, Bureau of Indian Affairs ca. 1901.

The insufficient size of land allotments and the lack of farming knowledge doomed any chance the tribe had of maintaining a self-sufficient economy. To gain income, Kiowas began to lease their lands to white farmers and ranchers. This practice continues today; Kiowas who farm or ranch their own land are the exception. Landholdings have become increasingly fragmented through inheritance, and quarter-sections with 20 or 30 Kiowa owners are common.

In order to make a living, many Kiowas have been forced to move from rural allotment lands to towns and cities, both in the region and even farther afield, and today, about half of the tribe's more than 8,600 members live outside the former reservation lands, with about a quarter living out of state. Because the federal government dissolved the Kiowa reservation almost a century ago and because half the tribe no longer lives there, one might think that the Kiowas would place less importance on southeastern Oklahoma. Yet this is decidedly not the case. The region remains an important psychological anchor for them, an intrinsic part of who they are. They maintain an intense loyalty to this particular area, a loyalty that their white neighbors are often completely unaware of.

Material and Mythological Landscape Features

The material and mythological imprint of Kiowas on the landscape binds them emotionally to their homeland and allows them to identify with it. Among the material features that strengthen the Kiowas' attachment to place are arbors and mission churches. The more mythological features include Devil's Tower, Palo Duro Canyon, Fort Sill, and Rainy Mountain.

Arbors

Brush arbors, sometimes referred to as "Indian shade" or "Indian air conditioning," are found among many groups of Indians west of the Mississippi River (fig. 9.2). Their construction varies widely, and can be anything from lean-tos of poles and branches to elaborate structures made by bending willow trees down, tying the ends together, and interweaving tamarisk branches. Some were made so strong that goats could graze on top. Arbors provide a shady, vented area for sleeping and cooking in the summertime, and many families used to live exclusively in arbors during the hot months. Many of my

Fig. 9.2. A Kiowa woman named Good Eye constructs her summer arbor near Anadarko, Oklahoma, in 1899. Photograph by Annette Ross Hume. Source: Phillips Collection photograph 523, courtesy Western History Collections, University of Oklahoma.

interviewees described vivid childhood memories of peaceful, slow summers spent in arbors.

Although traditional arbors were (and are) made of brush, they can also be permanent outbuildings; their specific construction material is not as important as their function. Although far less ubiquitous today than in the past because of the introduction of air conditioning, Kiowas still build and use arbors. Usually located behind houses, they provide great relief from the scorching Oklahoma summer sun. More important, the building of an arbor today is a statement of identity. It is a way for Kiowas to feel that they are living, at least in part, in the way their ancestors did.

In addition to family arbors, large circular arbors that formed part of the medicine lodge structure of the Sun Dance were also an intrinsic part of Kiowa culture. They served as shade for spectators. Modern versions of these circular arbors exist as well. Like the old ones, they provide shade for spectators at tribal dances, where they encircle the dance ring. Because canvas tents could serve the same function and would be easier to construct, practicality clearly is not the guiding rule. The circular arbor is a visible and meaningful link between the old Sun Dance and modern tribal gatherings and lends a sense of continuity to the proceedings. While their practical necessity has declined, their symbolic importance has grown greatly in modern times.

Mission Churches

The first white missionaries among the Kiowas were Quakers, who arrived in Oklahoma in the late 1860s (Corwin 1968). They represented a new policy of Congress whereby the control of Indian agencies passed from the military to religious groups (Hagan 1990, 57). Though officially charged with bringing the gospel, missionaries in reality were tools of the government in their continuing attempts to "civilize" and calm the "savage" Plains tribes (Hagan 1990, 58). Although the government in 1878 replaced the Quakers with agents who were less inclined to peaceful methods of control, mission work was still encouraged.

Methodists began work among the Kiowas in the late 1870s (Vernon 1980–81). They became a significant force on the reservation in 1887, when J. J. Methvin established a church at Anadarko. At about the same time, Catholic missionaries also became active, establishing St. Patrick's Church and an associated mission school at Anadarko. The Catholic school sparked Methvin to start a mission school of his own, lest too many souls slip into the wrong hands. Neither Catholics nor Protestants accepted the legitimacy of the other during these years (Vernon 1980–81, 401). Methvin's school was influential and instilled white values as well as the English language in the Indians in order to make the saving of their souls a less strenuous task. The school also trained significant numbers of Kiowa preachers and other religious leaders who would later spread the church's influence among the tribe. The Methodist mission church established at Mount Scott in the 1890s was an outgrowth of Methvin's work in Anadarko and became known as the "Mother Church of Kiowa Methodism" (Vernon 1980–81, 407).

Baptist missionary work among the Kiowas began in the late 1880s (Corwin 1968, RMKIBC 1993), and, at the invitation of Lone Wolf in 1892, missionaries were sent to the Kiowa camp on Elk Creek, south of Hobart. Despite protests from many Kiowas, the government gave the Baptists 80 acres near Rainy Mountain and another 80 acres near Elk Creek, along with five additional acres at each location for tribal cemeteries. Missionaries founded churches at both sites in 1893, and their successes led them to found other Baptist churches at Redstone and Saddle Mountain. Several generations after the death of Isabella Crawford, the founder of the Saddle Mountain Church in 1903, the Kiowas still speak of her with great reverence, as they do of the other missionaries who started Kiowa churches. Crawford's body is buried in Saddle Mountain Cemetery.

In the competition for converts, the Baptists and Methodists were much more successful than the Catholics, Quakers, Presbyterians, and other missions, and they remain the two principal Christian churches among the Kiowas today. Original Baptist mission churches are still in operation at Rainy Mountain, Elk Creek, Redstone, and Saddle Mountain, while the Methodists hold forth at Mt. Scott, Cedar Creek (east of Carnegie), Methvin (about 10 miles south-southeast of Ft. Cobb) and Botone (between Carnegie and Ft. Cobb) (map 9.1).

The government's goal in promoting mission activity was to assimilate, not to provide foci for, the Kiowa community. Ironically, however, the mission churches throughout Kiowa Country became and remain just that—centers of activity that set the Kiowa community apart from the surrounding non-Indian population. Although some Kiowas attend white churches, most who live in the area prefer the old missions. Part of the reason for this is a feeling of being unwelcome in white churches in the area, a manifestation of the subtle (and sometimes not so subtle) racism that many of my interviewees felt in Kiowa Country. Equally important, however, is the opportunity that the Kiowa churches allow for participation in a uniquely Kiowa brand of Christianity. The all–Kiowa language service of earlier years has disappeared, but much singing is still in the native tongue. Along with a few traditional English-language hymns that have been transcribed into Kiowa, many Christian songs unique to the Kiowas have been (and continue to be) written by church members, utilizing traditional Kiowa melodies.

Like many rural churches throughout the United States, the Kiowa churches are now experiencing declining attendance and membership. Despite the decline, however, the mission churches remain an important and vital force among the Kiowas. Whether their original goal of assimilation was successful, the churches did have the effect of tying the Kiowas to place by providing a concrete focus of activity for many communities. At the centennial celebration of the Rainy Mountain Church in the summer of 1993, one of the speakers addressed this issue: "Today as our people have sought education, there are doctors and lawyers among the Rainy Mountain People. They may live far away, but they have not forgotten Rainy Mountain, for their roots are here. Rainy Mountain People are those people all over the country who return as often as they can, because their memories are here. They bring their loved ones here for burial."

As Kiowas increasingly move away from their allotments to towns and cities, the rural locations of these churches make little practical sense. Many of the members of Rainy Mountain Church ("the Kiowa Vatican," in the words of one Kiowa) have to travel up to forty miles to attend church. But when given the chance to move the church to Mountain View, the congregation voted to keep it in its old location. Part of the reason is the natural beauty of the setting; from the church, few other human structures can be seen, and the Wichita Mountains stand silently on the horizon. Part of the reason is also the symbolic importance of the old location. There their ancestors chose the spiritual path that provided them with hope in the desperate times during the death of their traditional buffalo culture. There these ancestors are buried, and there their descendants wish to continue to follow what they still call the "Jesus Road."

Devil's Tower

Both in the physical landscape of the northern Plains and in the mythological landscape of the Kiowas, few landmarks are as prominent as Devil's Tower, in eastern Wyoming. Prominent Kiowa author N. Scott Momaday has recounted the legend, one shared with several other Plains tribes:

> There are things in nature that engender an awful quiet in the heart of man; Devil's Tower is one of them. Two centuries ago, because they could not do otherwise, the Kiowas made a legend at the base of the rock. My grandmother said: *Eight children were there at play, seven sisters and their brother. Suddenly the boy was struck dumb; he trembled and began to run upon his hands and feet. His fingers became claws, and his body was covered with fur. Directly there was a bear where the boy had been. The sisters*

were terrified; they ran, and the bear after them. They came to the stump of a great tree, and the tree spoke to them. It bade them climb upon it, and as they did so, it began to rise into the air. The bear came to kill them, but they were just beyond its reach. It reared against the tree and scored the bark all around with its claws. The seven sisters were borne into the sky, and they became the stars of the Big Dipper (1969, 8).

This story, which has its beginnings in the time when the Kiowas first came onto the Plains, does much more than simply provide a fanciful explanation for the formation of a prominent physical feature; it makes the tower and the stars landmarks of protection and rebirth. The talking tree stump becomes a metaphorical gateway from dog and travois life in the mountain wilderness to the horse and buffalo culture of the Plains. It is also a permanent representation on the landscape of the spirit forces that protected the Kiowas from the bear (a complex, sometimes ambiguous symbol throughout Kiowa mythology of evil and destruction as well as of power and strength) during and after the transition to Plains life (Boyd 1981, 10). Most of my interviewees viewed this relatively short period of equestrian culture as the "golden age" of their people, a time when the tribe was living closest to its "truest" self. In contrast, the earlier period in the mountain wilderness is seen as a constant struggle for survival, and the later period on the reservation as the dying gasps of this age.

Modern Kiowas still hold Devil's Tower to be an important part of their identity. The Kiowa Elders' Center in Carnegie, for example, plans trips to the monument almost every year. Even if an individual has never seen the Tree Rock firsthand, its image is still important: "I've never visited there in person," one woman told me, "but I've visited through the stories of our people; I think of it as mine."

Palo Duro Canyon

Palo Duro Canyon, in the Llano Estacado of the Texas Panhandle, has a darker relevance for the Kiowa. In what was to be the last gasp of armed Kiowa resistance to the sedentary reservation life imposed on them directly by the U.S. military and indirectly by the destruction of the vast buffalo herds of the Plains by white hunters, a group of Kiowas under the leadership of Lone Wolf and Maman'-te' attacked the Wichita Agency at Anadarko in 1874. Following the uprising, they and their followers, over half of the tribe's estimated 1,700 members, fled up the Washita River and into the Texas panhandle (Nye 1942, 210; Mooney 1979, 236). Experiencing miserable weather all the way, they finally took shelter in the canyon along with bands of Comanches and some Cheyennes who were also opposed to reservation life (Mayhall 1971, 295–96; Boyd 1983, 247–49).

On 17 September 1874, Colonel Ranald Mackenzie's forces found the encampment. Only about three Indians were killed in the battle, but Mackenzie captured 1,400 ponies, slaughtered most of them, and destroyed all of the Indian villages and property. Some of the survivors attempted to remain on the Plains through the harsh winter that followed, but without horses, lodges, or buffalo, it was impossible. Not only were horses the means of transportation for the tribe, they were symbols of wealth; their loss was singularly devastating.

The importance of the canyon in Kiowa history became clear only in retrospect, however. The annual calendars kept by the Kiowas scarcely mention the entire 1874 outbreak of hostilities (Mooney 1979, 145), but now the canyon has been accorded symbolic status. Although a few isolated battles followed, Palo Duro Canyon today has become a central symbol of the final military defeat of the Kiowas. Like Devil's Tower, the canyon has become a pilgrimage site that Kiowas often visit to connect with their past, and the tribe periodically organizes group trips.

There was little bitterness connected with the defeat at Palo Duro among the people I talked with, only a very palpable sadness. As one woman described her visit: "When we went down there, I wasn't really into visiting museums and things like that, I've only recently become interested in that. But when we were in there, coming by the bluff, I got a feeling and I began to cry. How frightened they must have been running around there trying to find someplace to hide. I could see them running around; I *felt* their fear." Inherent in every ending, however, is also a beginning, and through the process of enduring change, continuity can be achieved.

Fort Sill

Fort Sill, like Palo Duro Canyon, is another grim reminder to the Kiowa people of their military defeat. Founded in 1869 by General Philip Sheridan during his winter campaign to subjugate the southern Plains tribes (Nye 1942),

Fort Sill today is a massive training center for field artillery. Alongside the fort's modern warmaking equipment, much of the Old Post has been preserved and is open to visitors. Among the buildings is the "Old Stone Corral," built to thwart horse raids by the Kiowas, Comanches, and Plains Apaches. The commemorative plaque labels it as the final "roundup" point for those tribes. But what to most visitors is simply another monument to the colorful, cowboys-and-Indians movie history of the West is to the Kiowas a solemn reminder of the final military defeat of their people. As bands of Kiowas and Comanches straggled in to Fort Sill and surrendered after the battle in Palo Duro Canyon, they were imprisoned in the corral.

Many present-day Kiowas knew their grandparents or great-grandparents who were held captive there. One woman I spoke with asked if I noticed her light-colored hair. She explained it this way: After the surrender at Palo Duro Canyon, "at the stockade at Fort Sill, they kept the adults in the jail, but they let the very young and the very old stay in the camps, and they assigned a man to 'guard' them. My mother's mother's mother was raped by a guard at Fort Sill; he also raped a number of other 12-, 13-year-olds. That's where my white blood comes from."

Momaday has written of the corral as well: "My grandmother was spared the humiliation of those high gray walls by eight or ten years, but she must have known from birth the affliction of defeat, the dark brooding of old warriors" (1969, 6). The Old Post Guardhouse that held many Kiowa leaders after the surrender is also still standing. Today it is known as the Geronimo Guardhouse, after one of its most famous prisoners. One woman I spoke with told me, "I was very young when I first visited the Fort and saw the jail cells, but I could feel—that place has some feeling to it. People locked up and the people you're supposed to take care of still outside, out there." While Devil's Tower in Wyoming marks the beginning of the Kiowa "golden age," the prison and stone corral at Fort Sill represent all the forces that brought it to an end.

The "chieftain section" of the Fort Sill Post Cemetery is another important Fort Sill landmark for the Kiowas. Established in 1869, this cemetery was the only one in southwestern Oklahoma until Indian mission cemeteries were started in the 1880s. Fort Sill contains the graves of many of the most revered Kiowa leaders of the treaty period: Sitting Bear, Satanta, Stumbling Bear, Kicking Bird, Big Bow, and Hunting Horse of the Kiowas, as well as Quanah Parker, the "Last Chief of the Comanches." The inscriptions on the stones make clear what qualified an Indian for interment. Beyond the personal names, most of the stones list only tribal associations and whether the person signed either the Little Arkansas or the Medicine Lodge treaties (which gave up vast tracts of land in exchange for peace and annuities). A few stones also identify Indian scouts for the U.S. Army.

Originally, many of these Kiowas were buried there against the wishes of their families. Nevertheless, this cemetery too has become part of the homeland of the Kiowas. In 1963, descendants of Satanta (with the permission of the Texas legislature) moved his body to Fort Sill from the cemetery of the Huntsville Penitentiary where he committed suicide nearly 90 years earlier. A person involved in the repatriation told me that "it was kind of sad in a way to have it done, but we all wanted his bones to be back here home with us."

Rainy Mountain

Located just at the edge of the Wichita Mountains is a small, round-topped knoll. It would be unremarkable if it were nestled among the larger peaks of the range, but standing alone just beyond the edge of the mountains, it draws one's eye (fig. 9.1).

More than any other landmark, it serves as a symbol of the Kiowa people; indeed, Momaday titled his search for his own Kiowa identity *The Way to Rainy Mountain* (1969). On the Kiowa tribal logo, designed by Roland Whitehorse, Rainy Mountain is depicted on the warrior's shield, a representation of the "ancient Kiowa burial ground at the end of the great tribal journey" (Boyd 1983, 304). Of the area centering on Rainy Mountain, Momaday has written: "To look upon that landscape in the early morning, with the sun at your back, is to lose the sense of proportion. Your imagination comes to life, and this, you think, is where Creation was begun" (1969, 5). Elsewhere, he refers to it as "the center of the world, the sacred ground of sacred grounds" (1989, 244). Most of the Kiowas I spoke with about the mountain repeatedly applied the phrase "sacred" to it as well.

To understand the significance placed by the Kiowas on Rainy Mountain, it is important to understand the view tribal members take of their people's migration from the northern mountains. By and large, they do not see it as a response to military pressure from the Sioux and Cheyennes or to changing patterns in buffalo migration. Rather, it was a long-term journey that, from the start, had a purpose, a final destination. After much movement, they found a spiritual center for their activity in the Wichita Mountains, where they could bury their dead and return for generations. The base of this mountain, symbolically at least, marks the spot where the Kiowas realized that their southward journey was complete.

One of the early centers of Kiowa activity in the southern Plains developed around Rainy Mountain. This focus continued in the reservation era; "if there were a 'capital' of early reservation Kiowa life," one man said, "this was it." Today, even though many rural Kiowas have moved to towns and cities, the sacred mountain still triggers many strong and complex feelings.

Oklahoma's Kiowa Homeland

As evocative as many individual landscape features and landmarks are for the Kiowa, the importance of the homeland as an anchor for identity goes beyond this. No matter how widely dispersed the Kiowas become in search of economic opportunity, many never lose their desire to return to southwestern Oklahoma permanently and be buried in the land of their ancestors. One woman described this connection: "You see, Indian people, we don't ever consider that we leave home. We may not be there in body, but if nothing else, we're there in spirit." "Home," for her and for most other Kiowas I spoke to, refers to southwestern Oklahoma, a place to which Kiowas can always return, where they feel themselves a part of the life of the tribe, a place where they can find strength and restoration for their identity. This feeling goes beyond a collection of material landscape features and symbolic landmarks. It is a sense of wholeness, group identity, belonging, and community that can only be found in one region. The existence of this sense among a people, more than any other factor, makes a particular area their homeland.

Homeplaces

Part of the explanation of the Kiowa attachment to place can be found in landholdings. The act of land ownership, of having a homeplace, ties individuals to the region in a concrete fashion. Most land owned by Kiowas has been passed through generations of their family from the time that the reservation was dissolved and allotted in 1901. If one is inclined to draw a boundary around the Kiowa homeland, the area of allotments is as good a way as any to do so, for it roughly encompasses the area that my informants referred to as "Kiowa Country" (map 9.1).

Allotted land that remains in Indian hands is still held in trust by the Bureau of Indian Affairs; all sales or leases must be administered through and approved by this federal agency. Currently, the Bureau administers 282,599 acres of land for the Kiowa, Comanche, and Plains Apache tribes. The tribes as a whole own some of the land, but most of it is owned by individuals. Much of the trust land has been alienated through the years, and the process of fragmentation through inheritance makes it difficult for individuals to live off the lease money, much less make a living farming or ranching the land themselves.

Despite its increasing marginality as a means of making a living, trust land is still of great psychological importance to the Kiowas. As a number of people put it, it means that they will "always have a place to go back to." Even small parcels of land, fragmented through multiple generations, are important to the Kiowa sense of rootedness. More than pieces of land, they serve as genealogical reference points; they are the *specific* ground where one's forbearers finally gave up nomadic life and settled. One man told me about the allotment he currently lives on: "I've got a beautiful spring creek running where I live. It was my great-grandpa's land allotted in 1903 or 1904, and when he went there, he found this creek and the fresh spring water. It's really a big, cold, fresh underground river. Now I'm the great grandson, and I'm living there, I'm drinking it." Another Kiowa told me of family vacations to their old homeplace. Although the original house was by this time unoccupied and quite run down, they would sleep on the floor and bathe in the creek that ran nearby. She also told of an elder Kiowa who returns to his homeplace once each year. "He would drive to the old homeplace, there was no longer any house there, and he stayed and slept in the car for the weekend. Of course, we all told him that he could have stayed with us, but he wanted to stay at the homeplace. I don't want to say that it's a cleansing . . . maybe a reconnecting to the past. After it, he said that he felt better, really good."

Home

The psychological restorative power of the homeland is a subject that came up in my interviews repeatedly, and seems to me to be the *key* aspect of a homeland. Kiowas are constantly immersed in a white-dominated society, and many I spoke with often feel their tribal identity and values fraying with this contact. Periodically returning to southwest Oklahoma provides a means of restoring and sustaining a Kiowa's identity.

No matter how widely dispersed they may be in search of economic opportunity, many Kiowas never lose the desire to return to Oklahoma permanently. Although I rarely found older Kiowas in southwestern Oklahoma who had not spent a portion of their life outside the homeland to earn a living, I equally rarely found one who had never consciously planned to return from the very time he/she left. "You [whites] all go down there to Florida when you retire. Not us—we come right back here." Many Kiowas never feel completely at home in the environment of white, urban America. One woman who spent several decades in Midwest City, Oklahoma, told me about her experience: "There were lots of people around, but it was lonely." Another man, now living in central Oklahoma, related the following story:

> My generation, the baby-boom generation, many of us are still maintaining traditional values. It's difficult when you're surrounded by non-Indian values. I can practice the traditional ways, but nobody will understand, it will be just me. To get focused, you go back home. While I'm here, I try to keep it up, by reading, by visiting, by eating, but it's hard. . . . I used to be able to speak a little Kiowa, but being away from it for so long, it's been washed out of me. When I go home to the various ceremonies, I ask the elders why they are doing them that way, and they tell me that "I can't tell you as well as I could in Kiowa." This affects the emotional and psychological mentality of those 40 and over who don't participate every day, who are out of that area trying to make a living. The only way to keep that is to come back. The further away you get from the heart, the harder it gets.

Homecoming: The Kiowa Gourd Clan Ceremonials

This desire to return is clearly manifested in the annual Kiowa Gourd Clan Ceremonials. Held in Carnegie every 2–4 July, this is the single most important annual event for the Kiowa people. The Tiah-pah Society, a descendant of early tribal warrior societies, hosts the event that is nothing less than the modern equivalent of the Sun Dance, the annual religious event that formerly brought all the disparate bands of the Kiowas together in one place. The time of year corresponds roughly to that of the old ceremony. Tribal members come back from all over the country, planning their vacations around the event; in 1993, a quick glance at the license plates in the park one afternoon told me that people had travelled to this event from Arizona, New Mexico, California, Iowa, Texas, Colorado, Kansas, and the Oglala Sioux Reservation, as well as from counties all over Oklahoma. Moreover, the Gourd Dance is truly a Kiowa event for Kiowas (Ellis 1990); the uninformed gawker would be quickly bored by the absence of the stereotypical fast and flashy "Indian" dancing seen at tourist-oriented events. Grass dancing is held in the morning each day, gourd dancing and giveaways during the afternoon, and gourd dancing, intertribal war dancing, and giveaways into the evening.

Often, the people I spoke with used the term "pilgrimage" to describe their trek to Carnegie every July, and their use of this term is instructive. This word usually refers to a journey to a shrine or sacred place, and many who live away from the homeland truly do see their trek in this way. What draws Kiowas to this event every summer is more than the desire to watch people dance. Although the Kiowas also hold descendants' gatherings where family history and traditions are passed on, this is the only event bringing together all the family groups in the tribe. It is a homecoming, often the only chance that people who live far away from the region have to see each other. It is also a time of renewal, a time to reaffirm one's Kiowaness. As one man put it, it is a chance for the tribe to gather and say "here we are again, we are still alive, we survive. . . . Going back to southwest Oklahoma is important, there's a lot of memories tied to that place. I know all the sounds and the smells and the singing and the dancing from the July fourth celebration. It's all a part of who you are, it restores your feeling, your spirit, your place."

For those who return home each year, it is a restorative event, a chance to heal a Kiowa identity that may be feeling frayed by its immersion in white society. As another man told me, "I've been coming back here a long, long time, and these are the things that are real."

CHAPTER TEN

The Highland-Hispano Homeland

· · · · · · · · · · · · · · · · · · ·

Richard L. Nostrand

The quincentennial celebration of Christopher Columbus's encounter with the New World in 1992 reminded many in the United States of the major role Spaniards played in our history. Spaniards initiated the permanent European colonization of the United States—in Florida in 1565 and in New Mexico in 1598. In the 1700s they added present-day Arizona, Texas, Louisiana, and California to Spain's colonial empire. Events in the 1800s forced Spain to relinquish political control of her northern frontier between Florida and California, yet her people and their landscape impress remained. And in the twentieth century additional waves of Spanish-speakers further Hispanicized the United States: Mexicans immigrated to the American Southwest and beyond after the Mexican Revolution of 1910, Puerto Ricans arrived in metropolitan New York after World War II, Cubans went to Florida after Fidel Castro's revolution in 1959, and Latin Americans from a dozen or more countries headed for a variety of destinations beginning in the 1980s. Those who came in the twentieth century were leaving homelands, not creating them. Thus, homelands among Hispanic peoples in the United States derive from Spanish colonial times, and in this volume Daniel D. Arreola (chap. 7) and I argue their development among Tejanos in South Texas and Hispanos in New Mexico.

The five frames in map 10.1 summarize how the Highland-Hispano homeland evolved. When Spaniards reached New Mexico in 1598, Franciscan friars and soldier-settlers moved right into the villages of the Pueblo In-

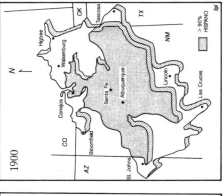

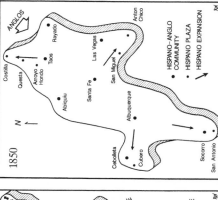

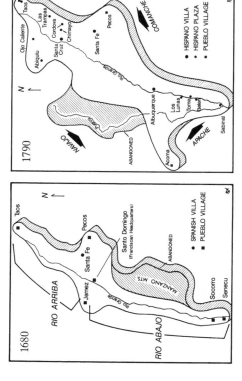

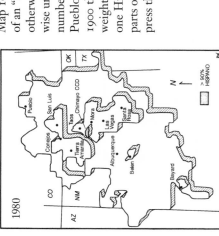

Map 10.1. The evolution of the Highland-Hispano homeland can be likened to the lifting up and partial collapse of an "island" on the land. In 1680 some 20 Spanish mission-conventos and Santa Fe formed enclaves in a low and otherwise uniform Pueblo Indian island. By 1790, 20 Pueblo villages now constituted enclaves in a low and otherwise uniform Hispano homeland island. By 1850, however, high Hispano percentages bolstered by larger Hispano numbers made the Hispano island stand more like a plateau, yet deep local depressions occurred at each of the 20 Pueblo villages (not shown) and at each of the eight large Hispano communities where Anglos had intruded. By 1900 the inner half of a now much enlarged Hispano island stood as a lofty plateau, but the island's outer half, weighted down as it were by Anglos, now formed a lower plain. And by 1980 the steady arrival of Anglos left only one Hispano erosional remnant, the Chimayo Census County Division. Below it Hispanos formed a majority in parts of eleven counties, and below that the weight of Mexican Americans had been added to Anglos to further depress the island's outer plain. *Sources:* see text.

dians. In 1680, when the Pueblos revolted, Spaniards in the Pueblo villages and in Santa Fe formed enclaves in a Pueblo Indian realm (frame 1680). Successful in their revolt, the Pueblos drove the Spaniards south to the Paso del Norte district, but in 1693 the Spaniards returned in a permanent way, and by 1790 they had transformed the Pueblo Indian realm into what D. W. Meinig (1971, 92) called a "Hispano Stronghold" wherein the Pueblos' villages were now the enclaves (frame 1790). When Anglos began to arrive in 1821, they created their own enclaves in Santa Fe and in other Hispano communities, but in 1850 the 55,000 Hispanos still constituted about 90 percent of the total homeland population (frame 1850). The Anglo intrusion continued, reducing the 140,000 Hispanos in 1900 to but 64 percent of their own region's population. By 1900, when the homeland reached about its full areal extent, Hispanos still constituted more than 90 percent of the population in half the region (frame 1900). But the relentless Anglo onslaught continued, augmented now by the arrival of Mexican immigrants, and by 1980 Hispanos exceeded 90 percent only in the small Census County Division of Chimayo. In the entire homeland, 365,000 Hispanos represented only 20 percent of the population (frame 1980).

As the proportion of Hispanos rose and fell over the four centuries, empirical evidence suggests that a positive correlation existed between areas of higher Hispano population and the degree to which Hispanos held a concept of their homeland. That concept, although abstract and elusive, concerns bonding to place. It has three basic elements: a people who lived in New Mexico long enough to have adjusted to its natural environment; a people who left their impress on that natural environment in the form of a cultural landscape; and a people who, from their interactions with the natural and cultural totality of the place, developed an identity with the land—emotional feelings of attachment, desires to possess, even compulsions to defend.

Environmental Adjustment

Consider first adjustment to New Mexico's environment. The natural environment that Hispanos bonded to is largely semiarid, meaning that it receives an average of only 10 to 20 inches of precipitation annually. Dry conditions explain the relatively sparse vegetation cover. The natural environment is also a highland with elevations that reach from 4,500 feet above sea level some 20 miles downstream from Socorro to over 8,000 feet at villages like Truchas, located midway between Santa Fe and Taos. High elevations in New Mexico mean rugged land and cold temperatures. Following the rising contours are basically two biotic life zones: the Upper Sonoran (4,500–6,500 feet), with its junipers and piñon pines, and the Transition zone (6,500–8,000 feet), with its ponderosa pines. From the outset Hispanos differentiated between the lower and warmer Rio Abajo, essentially a long swath of gently sloping Rio Grande floodplain where they found water and arable land to be plentiful, and the higher and colder Rio Arriba, where they discovered long stretches of the Rio Grande and its tributary valleys to be deeply entrenched and to lack arable floodplain (map 10.1, frame 1680).

That the homeland is semiarid and sparsely vegetated seems not to have elicited much reaction from Spanish and Mexican chroniclers of New Mexico. Both Spain and New Spain, after all, were basically dry places. That the Rio Arriba part of the homeland is high and cold, however, did evoke reaction. From Francisco de Coronado's explorations between 1540 and 1542 to Antonio Barreiro's (1928) description of New Mexico in 1832, the cold winter, especially in northern New Mexico, was the climatic element that most frequently prompted comment (Tuan and Everard 1964, 270–71). The cold temperatures, Spaniards understood, resulted from high altitudes; neither Tejanos nor Californios, who lived at about the same latitude as New Mexico but on lowlands near coasts, experienced such harsh winters. Significantly, Hispanos who lived in the Rio Arriba apparently perceived their rugged terrain to be preferable. In her ethnography of Hot Springs (now Montezuma), a small village set at the foot of the snow-capped Sangre de Cristo Mountains a half-dozen miles northwest of Las Vegas, Helen Zunser (1935, 143) noted in the 1930s that Hispanos "spoke of low flat land with derision" (fig. 10.1).

In adjusting to their natural environment, Hispanos encountered no major problems. To be sure, the severe winters in the Rio Arriba forced them to construct livestock shelters in addition to open corrals. While traveling through the Taos area in the dead of winter in 1846–47, George F. Ruxton (1848, 86, 91) was able to stable his mules in warm "sheds," where he fed them corn and corn shucks. And in the high Rio Arriba, where the growing season is short, wheat, a cold-weather grain, sometimes replaced corn. This tradeoff probably explains why people in the Rio Arriba prefer wheat tortillas today, while their neighbors in the Rio Abajo prefer corn tortillas. But in response to

Fig. 10.1. Adobe houses in El Cerrito, New Mexico. Spanish Americans or Hispanos founded villages that stretched from 4,500 to 8,000 feet above sea level. High elevations in the middle latitudes meant cold winters, as seen in the village of El Cerrito (elevation 5,650). Even at elevations above about 6,500 feet, where ponderosa pine was available for building, Hispanos preferred building with adobe brick, a material that insulated houses from the cold of winter and the heat of summer. Photograph by Richard L. Nostrand, 29 March 1980.

rainfall that is scant and unreliable, Hispanos irrigated their crops, much as they had in Spain and New Spain. Lacking timber at elevations below 6,500 feet, they constructed buildings with adobe brick, much as they had in Spain and New Spain. And when low-elevation pastures dried up in summer months, they drove their animals to lusher forage at higher elevations, an old Spanish practice known universally as transhumance (Foster 1960, 38, 56, 62, 70).

Besides water, what is precious in New Mexico is the limited amount of arable floodplain. To apportion this floodplain so everyone had access to water, Hispanos employed long lots, an ingenious adjustment to a scarce resource that Alvar W. Carlson (1975, 53) postulated New Mexicans developed independently in the mid-1700s. Stretching between irrigation ditches and rivers, long lots are ribbon-like fields onto which farmers feed water at the ditch end and gravity drains it across the floodplain to the river end. In New Mexico, where Roman legal tradition has heirs inheriting land equally, long lots are known to be subdivided into the narrowest strips, yet all heirs still have access to irrigation water. At the same time, then, long lots are equitable and efficient, and they accommodate population growth. And where long lots back up against rivers, Hispanos were careful not to disturb the trees that line the river banks. Louis B. Sporleder (1933, 29) reported that along the Cucharas River at La Plaza de los Leones (Walsenburg, Colorado), flooding happened only after Anglos cut down this riparian vegetation in the 1870s (fig. 10.2).

Destroying the riparian vegetation points up a contrast between the way Anglos and Hispanos went about adjusting to the same environment. The Harvard University Values Study characterized the difference as Anglo "mastery over nature" as opposed to Hispano "subjugation to nature" (Kluckhohn and Strodtbeck 1961, 13, 179). Subjugation means literally "to bring under the yoke of." Assuming that this is an apt characterization, Hispanos were more under the yoke of nature in the Rio Arriba than in the Rio Abajo. Rugged terrain, constricted floodplains, rigorous winters, and a short growing season clearly presented greater challenges for *arribeños* than for *rivajeños*. Not surprisingly, at the time of the American takeover the Rio Abajo was the more highly cultivated and productive of the two subregions (Sunseri 1973, 331–32). When writing of agriculture in 1844, Josiah Gregg (1844, 104 n. 9) depicted the Rio Abajo as containing "the principal wealth of New Mexico." But to say that nature "subjugated" Hispanos is perhaps overstating the case. The Hispanos' cultural baggage, after all, had prepared them for the dry environ-

Fig. 10.2. Long lots in the Pecos Valley near Villanueva, New Mexico. Water drains through furrows from the irrigation ditch (behind the pickup truck) to the Pecos River (lined by riparian vegetation) in the distance. Photograph by Richard L. Nostrand, 12 October 1985.

ment, and they had coped successfully with the high and cold Rio Arriba. Indeed, their excessive overstocking of the Rio Puerco Basin is an example of actual environmental exploitation.

The Landscape Impress

As Hispanos adjusted to their environment, they stamped it with their culture. There are hundreds of examples of the Hispano cultural impress, and although none seem to be unique, several make the Hispano homeland distinctive. Long lots are a case in point. Besides Hispanos, Tejanos employed long lots at San Antonio, Texas, in 1731 and along the lower Rio Grande in the 1760s. Noting that there is no precedent for long lots in Iberia or in New Spain, Terry G.

Jordan (1974, 70–74, 76, 82, 84, 86) postulated that the Spanish at San Antonio developed long lots independently and that those along the lower Rio Grande diffused from French Louisiana. He reported that Stephen F. Austin employed long lots in his colony in the Mexican era and that there are examples of the use of long lots in Texas after the Mexican era. But in Texas long lots are not commonplace. By contrast, in New Mexico they are ubiquitous. They dominate the cadastral pattern into which residents divide riverine properties. In a relative way they make the landscape of the homeland distinctive.

Villages are a second reason why, in a relative sense, the Hispano impress is distinctive. During most of their experience most Hispanos have lived in villages. In California and Texas, Spanish-speaking people also lived in villages. But in those two parts of the borderlands, especially in California during the Mexican era, land grants awarded to individuals for purposes of stock raising were, on a per capita basis, many times greater than those in New Mexico (Pitt 1966, 11). In California and Texas a greater percentage of people lived in dispersed ranchsteads. Thus, New Mexico, by far the most populous part of the borderlands in Spanish and Mexican times, also had the highest proportion of people living in villages. In recent decades urbanization has siphoned off many rural villagers. Yet New Mexico's "plazas," as their villages are known, survive. Sprawling affairs of remarkably low density, with houses dispersed in linear fashion along the high sides of irrigation ditches or at intervals along roads, these loose agglomerations of people dominate the landscape of the homeland. Indeed, in the American West, only Mormons, with roots in community-minded New England, live in rural villages to the same degree.

Log structures are a third example of the distinctive landscape impress. The Hispanos' habitat extends well above 6,500 feet into a ponderosa pine life zone, and although adobe brick is the most common building material, even at high elevations, Hispanos also build with logs (Gritzner 1974a, 26, 28, 29). Only in east Texas in the southwestern borderlands did Spaniards also have available to them a pine forest resource, but their small numbers precluded much use of it (Winberry 1975, 289, 292). In New Mexico, however, Hispanos commonly built houses, barns, outbuildings, and even structures to house gristmills of notched and interlocked horizontally laid logs (fig. 10.3) (Gritzner 1971, 54–62). There appears to be little doubt that Spaniards introduced the technology for log construction instead of Anglos, for the Spanish employed it in New Mexico as early as the middle of the eighteenth century (Gritzner

Fig. 10.3. Log barn in Tierra Amarilla, New Mexico. The logs in the first story of this barn (7,524 feet) sag under the weight of a framed second story. Logs are a common building material above about 6,500 feet. Photograph by Richard L. Nostrand, 19 October 1979.

1974b, 518, 519). And if John J. Winberry (1974, 54, 62–64) is correct that German miners took log construction techniques to Mexico's central plateau in 1536, then Juan de Oñate could conceivably have introduced this technology in 1598 to New Mexico.

It is certain that Oñate introduced the dome-shaped outdoor adobe ovens known as *hornos* (Ellis 1987, 30). Hornos are associated with wheat culture. They are found in Spain and in New Mexico and Argentina—the relatively cold opposite ends of Spain's New World empire where people grew wheat. According to Marc Simmons (personal communication, 1989), hornos are uncharacteristic of central Mexico, where corn is the favored grain. After their introduction to New Mexico, hornos and the paraphernalia associated with wheat culture, such as the metal sickle and milling technology, apparently diffused rather quickly among the Pueblo Indians. In Spanish and Pueblo villages, then, hornos became the standard oven for baking wheat (and corn) products. They are commonplace today. Given their scarcity in Mexico, as well as California and Texas, hornos are an item of material culture that sets the Hispano homeland apart.

The impress of religion is also distinctive. Every homeland village has its Roman Catholic church, centrally positioned with steeple and cross dominating the skyline. Villages with Penitente chapters have their *moradas*. In Chimayo the Santuario de Chimayo, which dates from 1816, annually draws thousands of pilgrims who seek to be cured by ingesting the *tierra bendita* or healing mud (de Borhegyi and Boyd 1956, 2–23). Along rural roads and in the countryside are a variety of religious shrines. Also dotting the homeland are scores of religious place names. Many, like San Miguel and San Jose, commemorate village patron saints (Chávez 1949), while others, like Santa Fe (Holy Faith) and Santa Cruz (Holy Cross), are simply religious terms (Chávez 1950). Except for the moradas and the Santuario de Chimayo, in a religious sense what differentiates the homeland from other sections of the borderlands is the sheer quantity of all these things.

Place Identity

Ordinary Hispanos know intimately every bump on the landscape and every turn in the road in their own *patria chica*, meaning their native village and its adjacent area. And like Spaniards in Spain, ordinary Hispanos have an intense love for their community of birth. Pride in their natal place is fierce and loyalty to it is unshakable, as in Spain (Foster 1960, 34–35). "To be Spanish American," wrote Margaret Mead (1955, 152) in her study of New Mexico, "is to be of a village." In New Mexico, the village of birth as much as the family name identifies an individual (Leonard and Loomis 1941, 8). What is known about the place identity of ordinary Hispanos, then, is their strong attachment to and identity with their own patria chica. Village prototypes in Spain explain why. A paucity of information, however, requires much conjecture when coming to grips with place identity beyond the patria chica.

Fray Angélico Chávez (1953) identified several images that formed the basis for a sense of place beyond the patria chica in the Spanish period. Until 1771, Spaniards called New Mexico "the Kingdom and Provinces of New Mexico." By "Kingdom" they meant Santa Fe, the other Spanish communities, and the Pueblo villages that resident friars staffed. Chávez noted that the people themselves referred to their land as "the Kingdom" and hardly ever as "New Mexico." He also recorded that the *villa* of Santa Fe, because of its size and importance, was usually called simply La Villa, its inhabitants *villeros*. Before 1771, at which time Spaniards created the *Provincias Internas* and all of New Mexico became a "province," the term *provinces* referred to the unsettled areas around the periphery of the kingdom, where the unchristianized Pueblo and nomad Indians lived. An eighteenth-century Spanish policy of establishing outposts, such as New Mexico's Ojo Caliente, Abiquiu, Las Trampas, Belen, Tome, and Sabinal, to serve as buffers against the nomad peoples, confirms that Spaniards perceived this periphery to be dangerous country. Indeed, in 1812, Pedro Bautista Pino (1942, 71) conceived of all of New Mexico as one giant buffer between the settled parts of New Spain to the south and the warlike nomad Indians to the north.

For the Mexican period David J. Weber (1985) likened geographical levels of the Hispanos' sense of place to widening circles of increasingly weaker loyalty (fig. 10.4). Hispanos had strongest loyalties to the innermost circle, the patria chica. At the next level came four *partidos*, or districts, whose head communities were Santa Fe, Santa Cruz, Albuquerque, and (until 1824) Paso del Norte (Ciudad Juárez, México). Then came the Rio Arriba and the Rio Abajo, subregions bifurcated by the so-called Rio del Norte, which Barreiro (1928, 79) referred to in 1832 as the "Nile" and "the soul of the territory." At the next level came New Mexico itself, officially a "territory" (1824–36), then a "department" (1836–46) in the Mexican era. Finally, Hispanos had an awareness of Mexico as a nation. Barreiro (1928, 73) wrote that in 1832 New Mexicans celebrated *el diez y seis de Septiembre*, yet loyalties to distant Mexico seem to have been less strong than they were to the home province, which was also the case in California (Pitt 1966, 4, 6, 7, 25, 53, 174, 309). Beyond the cordon of outposts that surrounded New Mexico the *indios bárbaros* constituted a dangerous frontier that Lansing B. Bloom (1913, 12, 34) said commenced just south of Socorro, which, as late as 1843, Hispanos referred to as *la tierra afuera*, the land outside.

The American takeover brought many changes in the Hispanos' sense of place. In *No Separate Refuge*, Sarah Deutsch (1987, 101, 108, 116, 126, 155, 163, 164) analyzed one of them. About 1870, Hispano men began to leave their villages to work seasonally for Anglos. Anglo pull was always strongest to the north, especially to the coal mines and sugar beet fields of Colorado. Entire

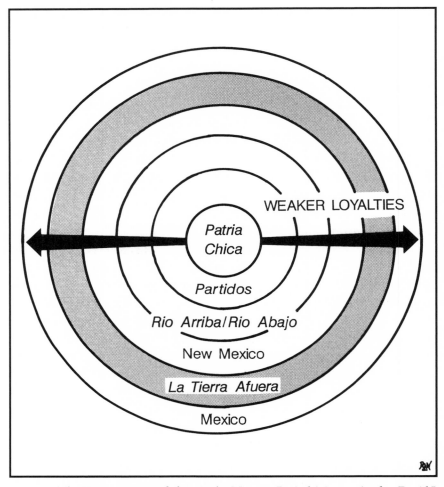

Fig. 10.4. The Hispano sense of place in the Mexican Period (1821–46), after David J. Weber (1985). Shown are widening geographical levels of weakening loyalty.

families eventually joined the migrant circuit, and many remained permanently. To use Deutsch's terms, "networks of kin" evolved that found thousands of Hispanos interacting between their northern New Mexico "village heartland" and "outposts" in Colorado—all in one giant "regional community." Probably no contemporary Hispano used these terms to describe the phenomenon, yet for the thousands who engaged in the seasonal pulsations out of New Mexico or who relocated permanently to coal mining camps and beet field *colonias*, a heightened awareness existed of the larger region beyond the patria chica. Wages earned in Colorado, moreover, flowed back to the villages, an affirmation of loyalty to the patria chica, and perhaps, as Deutsch argued, a Hispanic strategy to preserve a threatened culture. In the 1930s the "regional community" collapsed, thus ending Colorado's function as a "separate refuge."

Conclusion

So what is the Hispanos' concept of a homeland? The term in Spanish that comes closest to capturing the idea of a homeland is *patria*, which means fatherland. Patria embodies the aggregate of the hundreds of patrias chicas that ordinary Hispanos know intimately and for which they have sentimental and enduring feelings of attachment. The Hispanos' concept of a homeland is then the totality of the patrias chicas. Their concept encompasses how they adjusted to these environments, how they stamped them with their culture, and from both the natural environments and the cultural landscapes, how they created a sense of place. And for Hispanos the concept of a homeland embraces a level of territorial consciousness or place identity that today is uncommon in mainstream American society.

CHAPTER ELEVEN

The Navajo Homeland

. . . .

Stephen C. Jett

The Navajo ([*T'áá*] *Diné*['*é*], [Just] the People, or *Naabeehó* [*Diné'é*]) are an Apachean-speaking people, the large majority of whom currently reside on and near one major reservation and its smaller satellites in northeastern Arizona, northwestern New Mexico, and southeastern Utah. The area of Navajo occupance is over 24,000 square miles and exceeds the size of West Virginia. The most populous Native American group, numbering about 219,000 in 1990, the Navajo have by far the largest reservation and quasi-reservation population (143,405). In 1990, they comprised a third of all U.S. reservation Indians (Paisano 1993; Shumway and Jackson 1995, 187–89, 199), despite some 35 percent living outside these lands. Thus, spatio-demographically, the Navajo are the most important Indian "tribe" in North America north of Mexico. They also control vast resources, notably in the realm of energy-producing mineral wealth.

Unlike many tribes, especially eastern ones, the bulk of the Navajo population remains in its traditional area of occupance (map 11.1), it is the majority group there, to the point of near exclusivity almost everywhere, and it has retained its traditional culture to an unusual degree. These facts contribute to a highly developed sense of "homeland"—unequaled among non–Native American ethnic groups in the United States.

This strong sense of homeland exits even though history and archaeology indicate that the ancestors of the Navajo arrived relatively recently in the American Southwest (Perry 1991), an arrival not attested to before the fif-

Fig. 11.1. Shiprock, northwestern New Mexico, with Navajos in foreground. This sacred volcanic neck is seen as part of a gigantic avian figure and was the nesting site of the people-devouring Rock Eagle Monsters. The homeland was freed of these and other monsters through the efforts of Monster Slayer and Born For Water, twin offspring of the earth deity Changing Woman by the Sun Bearer. Photograph courtesy the Western History Department, Denver Public Library, author and date unknown.

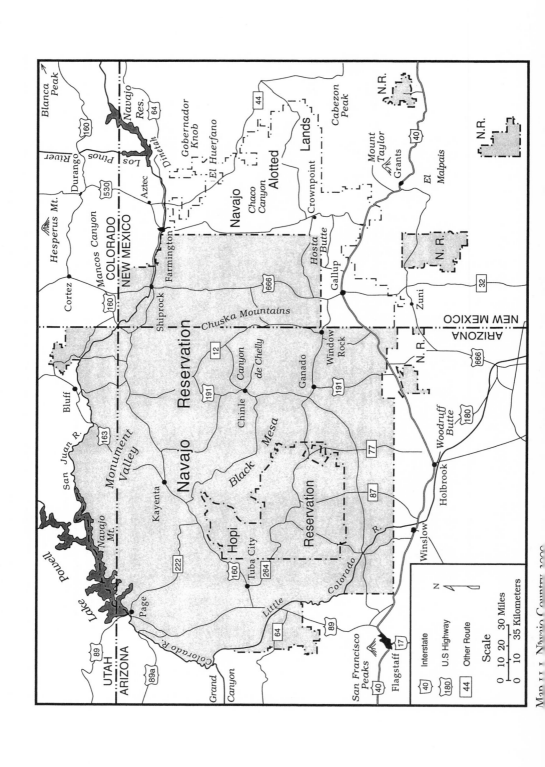

Map 11.1. Navajo Country, 2000.

teenth century. These "pre-Navajos" migrated southward after splitting off from their Northern Athapaskan-speaking kin in western Canada. Settling in the upper San Juan River drainage of northwestern New Mexico and adjacent Colorado on the edges of Pueblo Indian territory, these people from a northern hunter tradition adopted flood-water farming (stressing maize, squashes, and beans), aspects of religion (for example, the idea of directional sacred peaks), and a variety of items of material culture (including weaving) from the Puebloans. Subsequent to the arrival of the Spanish around 1600, these "proto-Navajos" borrowed livestock-raising, especially of goats, sheep, and horses, from the newcomers, via the Pueblos. Their domesticated animals increasingly replaced wild game as a source of meat. Acculturation went on from early times but was probably particularly intense in the period following the return in 1692 of the Spanish after the 1680 Pueblo Revolt, which had forced withdrawal of the latter for a dozen years. At this time, a considerable number of Rio Grande Puebloans fled to the Navajo Country to escape Spanish reprisals, and there seems to have been intermarriage (Brugge 1968), leading to a fusion into fully Navajo culture.

By 1750 a way of life had developed among the Navajo, that revolved around maize agriculture and pastoralism, supplemented by hunting and gathering. This lifeway continued to develop over time; the degree of emphasis on one or another of these activities varied locally according to environmental possibilities and constraints and temporally according to military pressures from outside groups (Brugge 1983). Settlement was in small, often isolated, extended-family homesteads (Jett 1978b, 1980) featuring hogans—conical (and later domical) earth-covered one-room dwellings (Jett 1992b). The people made seasonal moves, usually between summer lowland farms and higher-elevation winter sites where firewood was available, or between lowland and highland pasturage (Jett 1978a). A distinctive cultural landscape came to characterize the region: patches of farmland, mostly on floodplains; homesteads with hogans, brush-roofed summer shades, corrals, and beehive ovens; and conical sudatories (Jett and Spencer 1981).

Sacred Mountains, Sacred Land

The Navajos' Origin Myth "tells The People something about their place in the universe. . . . Man is understood as part of nature along with animals, insects, and features of the land; he is not seen as necessarily superior or destined to dominate" (Iverson 1981, xxx). "Thus all beings and aspects of nature are *atk'éí:* 'those who should be treated with compassion, cooperation, and unselfishness by the Navajo' (Witherspoon 1975:37)" (Schwarz 1997, 46).

"Navajo people are homologues of the Navajo universe" (Schwarz 1997, 47). Regarding the relationship between the Diné'é and the Earth—specifically, the Navajo earth, the homeland—we may begin with the following observation: "The land is at the heart of traditional Navajo religious beliefs and practices [*Diné binahagha'*]. The whole Earth is considered sacred. It yields plants, minerals, and other natural materials for religious observances. Certain places are where rituals are performed. Certain places are important in traditional [tribal and clan] origin histories. Religious observances, such as blessing of dwellings and cornfields, accompany many of the mundane practices" (Kelley and Whiteley 1989, 4). Further, "Nearly all aspects of the natural world are personified" (Beck and Walters 1977, 76) and may be communicated with. Navajo religion "must be seen as a design in harmony, a striving for rapport between man and every phase of nature, the earth and the waters under the earth, and the sky and the 'land beyond the sky,' and of course the earth and everything on and in it" (Gladys Reichard, in Kammer 1980, 19).

These attitudes and practices are inextricably bound up with cosmology. For the Navajo, "Earth and Sky are the two major and englobing cosmological phenomena; they constitute the couple defining the ultimate boundaries of the Navajo universe. . . . Earth and Sky are said to stretch out as anthropomorphic figures from the East (head) to the West (feet), the Sky lying on top of Earth like a man lying on top of a woman in the sexual act. . . . [but] without touching each other" (Pinxten and van Dooren 1992, 102). In the space between Earth and Sky, zoned vertically, are the stars, the moon, the sun, and the ambient air. The four cardinal directions have their four personified "inner forms": (White) Dawn Man in the east, Horizontal Blue Man in the South, Horizontal Yellow (Evening Twilight) Woman in the West, and Darkness Woman in the north, forming a circle (Haile 1943, 71).

"'We are told our legs are made from earth, our mid-section from water, our lungs from air and our head is made out of heat and is placed close to Father Sun. We are known as the On-Earth-Holy-People. For that reason, our skin is brown like Mother Earth. . . .' After a life cycle, people return to earth, the mother, [medicineman Benny] Silversmith says" (Reid 1992, 119). "The corpse, in decomposing, is to the benefit of the Earth, and therefore of everyone" (Farella 1984, 131).

All the Earth and creation (*Niilyáii*)—made from the elements moisture, air, substance, and heat and enlivened by vibration—is holy (Jett 1992a, 30; Kelley and Francis 1994, 100; Schwarz 1997, 46). More specifically, the region bounded by the four paramount sacred mountains is what was made expressly for the Earth-surface people (*nihokáá' diné'é*) by the Holy People (*Diné'é Diyin, Diyinii*) and is the supernaturally sanctioned homeland. The territory defined by these directional mountains is *the* world, exclusively for the *Diné'é* or People (usually given as *Diné*, the singular)—that is, the Navajos.

These mountains are (in ceremonial sunwise order): East (prime direction), Blanca Peak, Colorado (*Sisnaajinii*, Horizontal Black Belt; Sleight 1950); South, Mount Taylor, New Mexico (*Tsoodził*, Tongue Mountain; Blake 1999, 502–4); West, San Francisco Peaks, Arizona (*Dook'o'oosłííd*, It Has Never Melted and Run off from Its Summit); North, Hesperus Peak, Colorado (*Dibé Ntsaa*, Bighorn Sheep). In the middle, in New Mexico, are El Huerfano mesa (*Dził Ná'ooditii*, The-People-Move-Around-It Mountain), which is Earth's lungs, and Gobernador Knob (*Ch'óol'í'í*, meaning unclear; sometimes interpreted as the Spruce Hill), said by some to be Earth's heart; also important is Hosta Butte (*'Ak'i Dah Nást'ání*, One Thing Atop Another; Gold 1994, 42). Gobernador Knob "is sometimes considered as the center of the world" (Reichard 1963, 20). *Dibé Ntsaa* is the heart of the Earth's inner form, say some, *Tsoodził* its tongue, *Ch'óol'í'í* its mouth, and the Hogback (*Tsétaak'á*), near Shiprock, New Mexico, its diaphram, for breathing (Griffin-Pierce 1992, 70–72). Said a Navajo man:

> The white people all look to Government like we look to the Sacred Mountains. You, the white people hold out your hands to the Government. In accord with that (the Government) you live. But we look to our Sacred Mountains: To Sierra Blanca Peak, to Mount Taylor, to San Francisco Peak, to La Plata Mountain, to Huerfano, and to Gobernador Knob. According to them we live—they are our Washington. (Quoted in Young and Morgan 1954, 17)

Navajo Country

The original known area of proto-Navajo occupance in the Southwest, flanking the San Juan River in northwestern New Mexico and adjacent Colorado, is known to the Navajo as *Dinétah*. The term means "Among, or In the Area of, the [Navajo] People"—that is, Navajo Country. This is where the mythological pre-Navajo First People settled after the Emergence from the lower worlds. Some say the Emergence took place in the San Juan basin (Witherspoon and Peterson 1995, 34); others believe the First People emerged at Island Lake in Colorado's La Plata Mountains (Van Valkenburgh 1941, 86). The deity Changing Woman (created at Gobernador Knob by a conjunction between sky and earth) is believed by Navajos eventually to have traveled from Dinétah to the shores of the Pacific, and there to have made the people of the first true Navajo clans from scrapings of her skin, who then (without her) migrated back to the area of her creation. Historical tradition confirms Dinétah as the original Navajo homeland, and the earliest archaeological remains identifiable as proto-Navajo and Navajo are found there (Hester 1962). Early sites are particularly numerous in the Largo and Gobernador canyon drainages, which can be thought of as the former core area. A "greater Dinétah" is sometimes said to have extended from Farmington, New Mexico, to the Rio Grande, and from Blanca Peak, Colorado, to Chaco Canyon, New Mexico.

Although today Navajos inhabit only the southwestern fringes of core Dinétah, that region is still considered special, even a holy land, and there are certainly highly important sacred places within it (Roessel 1990), to which individual pilgrimages are still made (Martin 1995). Despite desecrations from a major reservoir and from gas and oil exploitation and other contemporary phenomena such as electronic communications towers atop El Huerfano (Jett 1995b, 42), Dinétah continues to be "considered by the Navajo to be special and sacred because it is their place of origin. . . . It, therefore, represents the sacred homeland of the Navajo" (Witherspoon and Peterson 1995, 34).

The Navajo Diaspora

Although some Navajos moved early into country to the west, Dinétah remained the core of Navajo occupance until the neighboring Ute in southwestern Colorado acquired firearms and forced the Navajo westward. About the mid-1700s, the Navajo abandoned most of old Dinétah and gradually (or, sometimes, quickly) spread toward the west and southwest, some ultimately reaching the Colorado River and eventually filling much of the then largely unoccupied country south of the San Juan River and north of the Little Colorado (Hester 1962). Perhaps it was at this time that the Navajo came to identify the present four Sacred Mountains as such, and to perceive the sanctity of Dinétah as including the entire area lying between those peaks.

Anthropologist Gladys Reichard (1963, 19) observed, "Traditionally, the Navajo tribe has always been on the move. They love to travel, yet feel a deep attachment to their present habitat. They have an extraordinary interest in geography. The number of places in myth and ritual is legion." One manifestation of this is the plethora of identifiable sacred places (*hodiyin* or *dahodiyinii*), associated with incidents in myth, including specific supernatural events as well as stops on tours of inspection by Holy People. Most of these are prominent landmarks such as mountains (seen, often, as the hogans of Holy People), pinnacles (some of which are considered to be petrified Holy People), and rivers (living entities that are viewed as having sexual congress at confluences with other streams); but many places, including small springs, that are sacred owing to mythic associations, are less physically obvious.

Each sacred place represents a concentration of the power that imbues all creation, power that, with proper knowledge, may be drawn upon for the good of the individual, the community, and the People as a whole. On the other hand, disrespect for, and desecration of, these places can cause the loss or misdirection of their power and result in disasters (Jett 1992a, 30, 36–37; Johnson 1987, 23). In addition to the sacred sites themselves, many are linked by traditional pilgrimage routes (Kelley and Francis 1994, 96).

As reflected in the myths, "the very act of traveling sanctifies" and empowers the traveler (Astrov 1950, 49). The itineraries of some of the Supernaturals and heroes reminds one of the ancient navigational *peripli* and *itineraria* of the Mediterranean world, and no doubt their reiteration helped many Navajos, especially medicinemen, to become familiar with the geography of their country and to be able to take long journeys even into areas in which they had never before been, without being significantly disoriented.

Navajo place-naming practices tell something about how they perceive their homeland and relate to it. Some sacred places have esoteric, often honorific, ritual names as well as secular ones. Almost all Navajo place names are translatable, and are very largely descriptive—for instance, Standing Black Rock, Dove Spring; very few are commemorative (in contrast to Euro-

American practice). Names of features other than family farms and modern settlements refer largely to natural landmarks—mountains, buttes, water bodies, and so forth, as well as to trails (Jett 1970, 180; 1997, 2001).

Exile from, and Return to, the Homeland

When the United States established its hegemony over the Southwest in 1846, Navajos and *Mexicanos* had a relationship of mutual raiding and slave-taking. The U.S. Army inherited this situation and attempted to resolve it. Efforts included several military expeditions into the Navajo Country and the signing of treaties (McNitt 1972; Brugge and Correll 1971). But owing to the withdrawal of most U.S. troops from the Southwest during the Civil War, many Indians, including Navajos, resumed raiding Hispano and Anglo settlements.

During a campaign in 1864–65, forces under the command of Col. Christopher "Kit" Carson succeeded in definitively defeating the Navajo (Trafzer 1982). The majority of the Navajo were then marched (via *Ninadá 'Iishjideedaa*, the Long Walk, of some 400 miles) to the Bosque Redondo Reservation at Fort Sumner in eastern New Mexico (although a considerable number remained at large, mainly in remote areas). The period of incarceration was one of considerable acculturation, accompanied by major hardship. After four years, the government acknowledged that the Bosque Redondo experiment had failed (Locke 1979; Moore 1994, 1–31).

The government then proffered the Navajo land in Oklahoma, but they rejected the proposal. Principal headman Herrero stated, "what we want is to be sent back to our own country. Even if we starve there, we will have no complaints to make" (Bailey 1964, 231). Anton C. Damon, one of the officials at the Bosque, recorded the following:

> [General William Tecumseh] Sherman wanted to send the Navajos to Indian Territory, where the soil, he said, was very fertile. Everything planted grew big and fine. They would be given land for farming, and seed and tools and everything.
>
> But the twelve Navajo Chiefs of the Council said they would not go to Indian Territory unless they were tied hand and foot and hauled there, and then they would run away. The Indians wanted to return to their former homes. . . .
>
> General Sherman . . . told the Chiefs they must have a fine country to want to go back to it so bad[ly]. Why did they like their country best?
>
> Then the Chiefs began to brag up their own country. When they planted wheat

there were two heads to every stalk. They planted potatoes as big as marbles and they grew as big as a man's head. The pin[y]on nuts grew so thick on the trees that they could fill a wagon in one spot. If he would let them go back to their own country they would never steal or kill people again.

. . . Sherman told the Indians he was going to send them back to their own country, but they must promise never to murder or steal again (Shinkle 1965, 70).

Besides simple familiarity and perceived (or deliberately exaggerated) productivity, as well as a greater distance from the enemy Comanches, one reason that the Navajos were particularly anxious to return to their homeland was the belief that their curing ceremonies were efficacious only within the Navajo Country—which was one perceived cause of people (and livestock) dying at Bosque Redondo. Headman Barboncito (*Hastiin Dághaa'í*, Mister the Whiskers) said, "we were told by our forefathers never to leave our own country" (Link 1968, 3).

Barboncito implied that the relationship between Navajoland and the Navajo people was a reciprocal one: "After we get back to our country it will brighten up again and the Navajos will be as happy as the land, black clouds will rise and there will be plenty of rain. Corn will grow in abundance and everything look happy" (Link 1968, 5, 9). This statement reflects the fact that "the Navajo view of the environment emphasizes participation and reciprocation" (Grinde and Johansen 1995, 107).

It was the Holy People who gave the country bounded by the sacred mountains to the Navajo people; by remaining within this homeland, the Navajo show their respect and appreciation for this precious gift of land. Within these boundaries, the efficacy of their ceremonies are [sic] ensured as well as their general prosperity. . . . Such a finely tuned sense of place means that being forced to leave one's homeland results in psychological trauma that is unimaginable to those of us without such geographic attachments (Griffin-Pierce 1992, 4–5).

General Sherman described the proposed new reservation to the Navajo leaders as including Canyon de Chelly and the Chuska and Carizzo mountains and the middle San Juan River but not Black Mesa. Barboncito responded that the leaders were "well satisfied with that reservation. It is the very heart of our country and is more than we ever expected to get." Headman Ganado Mucho (*Tótsohnii Hastiin*, Mister Bigwater Clansman) also expressed his pleasure. A treaty was signed.

Nevertheless, many Navajos apparently believed that the treaty guaranteed "that these Encircling Mountains would always be ours, so that we could live according to them" (Young and Morgan 1954, 17; also, Brugge and Correll 1971, 88–98), even though the Mountains were in fact excluded from the treaty reservation and remain outside the present Navajo lands. This impression may have been given by General Sherman's statement, during negotiations, that any Navajo could settle on any unoccupied land in the Territory (like other homesteaders), although they would then be subject to the same laws as non-Indians. In fact, conversion of the Navajo to pure farmers on 160-acre allotments would become federal policy (Moore 1994, 28–29).

Evolution of the Homeland Reservation

One Navajo headman, Manuelito (*Hastiin Ch'il Haajiní*, Mister Black-Plant Area), describing the return from Bosque Redondo, stated, "When we saw the top of the mountain [Taylor] from Albuquerque, we wondered if it was our mountain, and we felt like talking to the ground, we loved it so, and some of the old men and women cried with joy when they reached their homes" (Kammer 1980, 24).

Since that time, the term *Diné Bikeyah* has become the usual one for referring to the land to which the People have title and on which they reside. It means the "Navajo's Land." It is the term used to refer to today's legally designated Navajo-owned areas: the reservations (*Naabeehó Bináhásdzo*, Navajo's Area Marked Around), allotted trust lands, and tribally purchased lands. The history of the creation and gradual enlargement of the Diné Bikeyah, as well as the histories of political governance of the area and intertribal land disputes resulting in land loss, are very much part of the homeland story. That history is quite complex.

The Navajo Country had no formal boundaries until 1868. The treaty (signed on June 1, 1868) that ended confinement to the Bosque Redondo established title to a reservation that belonged to the tribe communally but was held in trust for the People by the federal government. The treaty reservation of 6,120 square miles was far smaller than the actual area of pre-1864 occupance. It included, as Barboncito said, the heart of the Navajo Country, but it also left out much land. The treaty accorded the tribe hunting rights, but not rights of occupance (other than through homesteading), in surrounding unoccupied lands. Nevertheless, most Navajos—not being familiar with parallels and meridians—probably simply assumed that their old territory had been restored to them and returned to their former lands wherever those might have been. Thus, from the beginning of the reservation period, Navajos occupied nonreservation lands, in violation of the treaty—no doubt sometimes knowingly and sometimes not. Navajo John Redhouse (1984, 9) wrote of the Indians' lack of understanding of the Euro-American concepts of land ownership and of their desperation to make a living where they could.

Within a decade officials recognized that the reservation of 1868 was too small to contain all Navajos; in fact, half lived off the reservation (Moore 1994, 202–4, 206). A long series of executive orders and legislative actions brought the Navajo Reservation and its satellites (the Ramah, Canoncito, and Alamo reserves) to their present configurations (Correll and Dehiya 1972). In 1974, after much legislation, litigation, and controversy, Congress ordered a large area that had been deemed joint property of the Navajo and Hopi tribes partitioned between them, resulting in the expulsion of thousands of Navajo residents from what is known as the former Joint Use Area in Arizona (Brugge 1994; Clemmer 1995, 232–72).

The Navajo Nation

American Indians became citizens of the United States in 1923. During the decades between the World Wars, the Federal Indian Service created a Navajo tribal government, and that government has continually increased its autonomy and asserted its residual sovereignty over Diné Bikeyah (Shepardson 1963; Young 1978).

In 1952, the Navajo government adopted its Great Seal of the Navajo Tribe and the Navajo flag. Both symbolize the land, showing the four Sacred Mountains, sacred plants, and livestock, surrounded by a protective rainbow. In 1969, the Navajo Tribal Council's Advisory Committee passed a resolution calling for use of the term *Navajo Nation* in place of the then official *Navajo Tribe;* the resoluton stated, in part, "It is becoming increasingly difficult for the Navajo People to retain their identity and independence, and it appears essential to the best interests of the Navajo People that a clear statement be made to remind Navajos that both the Navajo People and Navajo lands are, in fact, separate and distinct" (*Navajo Tribal Code,* quoted in Iverson 1981, xxv). Finally, in a governmental reorganization in 1989, the name *Navajo Nation* was officially adopted.

Levels of Homeland Attachment

The Navajos' sense of home and homeland exists on various levels. At the highest level is the sense of belonging to the Navajo Country as a whole. This sense was, until recent decades, perhaps mainly religious and abstract and associated with Navajoland itself and not involving a significant sense of membership in a "tribe." Within the Navajo Country are many local communities, descendants of old-time small local bands or land-use communities that occupied given areas (and since about 1930, usually organized into the Indian Service–imposed local political entities known as chapters; Bingham and Bingham 1987; Goodman 1982, 20–21).

Until after World War II (and to some extent to the present), these local communities—and, above all, the clan (*dóone'é*) and local extended-family kin groups (*dah 'oonéłígíí*) comprising each community—were more prominent in peoples' minds than the "tribe" (*diné'é*) or reservation or Navajo government as a whole. Within the communities are the territories (*diné t'áá bił nahaz'áagi*) of individual families (homestead groups) or extended families (resident lineages; Jett 1978b), usually called "customary-use areas" (Kelley 1986). There was and is a high level of attachment to one's customary-use area, from which the family makes or made a living, and, within that area, to the hogan (Jett 1995a). This attachment is symbolically reinforced by local ritual burial of a neonate's umbilical cord and by other practices (Schwarz 1997), a story told elsewhere (Jett 1998b).

Recent Developments

For many decades, Indian educational policy attempted to acculturate Indians to the dominant culture. From the 1880s to the 1930s, the government outlawed certain "offensive" Indian religious practices (Beck and Walters 1977, 157–64) and until the 1960s forbade the use of native languages at school (Lord 1996, 68). These policies have been largely reversed. Ceremonials, including formerly forbidden ones, are currently common in the Navajo Country, although certain ones have become extinct or nearly so. Beginning in the

1930s, and especially since the 1960s, day schools, including public (as opposed to Bureau of Indian Affairs) schools, have increasingly replaced Bureau of Indian Affairs boarding schools, and this, made possible by new school-bus roads—especially since 1960—allows children to live with their parents and receive cultural reinforcement.

Beginning tentatively in the 1930s (Thompson 1975, 56), but growing more prominent from the 1970s, Navajo language (*Naabeehó bizaad*) and culture have become a part of Bureau of Indian Affairs's curricula. Starting in 1966 with the demonstration school at Rough Rock, Arizona (Roessel 1977), about 16 schools instituted Navajo school boards and Navajo-oriented curricula and curricular materials (Frisbie 1992, 495–96; Pollack 1984, 172–73, 180; Iverson 1983, 636–37; Pavlik 1990; Lindig and Teiwes 1991, 197–201). Some involved traditional farming methods (Bingham and Bingham 1979), mythology, sacred places, and Navajo language.

Navajo Community College (now known as Diné Community College) was established in 1968 and has several branches. It offers instruction in traditional healing and has adopted the Diné Philosophy of Learning (DPL), in which the organizing principle is a balancing of the kinds of learning associated with each of the traditional four directions. The prime direction, east, is that of knowledge leading to sound decisions; the south, of earning a living; the west, of social well-being and human relationships; and the north, of respect for and reverence of nature (Frisbie 1992, 500–501; Emerson 1983, 669–70). This geographical organization of knowledge reinforces a sense of homeland.

Being rural and relatively remote until rather recently and in a region with a fairly sparse Euro-American population, the Navajo long remained somewhat freer of major outside acculturative forces than did other tribes. True, forced schooling, including off-reservation boarding schools where native language and culture were suppressed, damaged cultural continuity among many now-older Navajos. However, considerable numbers of children were incidentally or deliberately kept from attending school and carried on traditional ways.

World War II saw major alterations in the high degree of insulation enjoyed by the Diné, owing to Navajos' experiences in the military and in other vital off-reservation employment (Underhill 1956, 241–58). These changes accelerated with the Navajo-Hopi Rehabilitation Act of 1953 (Young 1961, vii,

1–5), roadway expansion beginning in the 1960s, Navajo families' acquisition of pickup trucks, burgeoning tourism (Jett 1990, 1998a), and mass communication—especially broadcast and video-cassette television—and other (coveted) impingements of modern life. Yet, despite rapid deculturation among many younger Navajos, native speech and folkways continue to predominate among tens of thousands of Navajos. It is an irony that although the percentage of Diné-speaking Navajos is declining rapidly—reduced to 10 percent among children entering school (Lord 1996, 68)—high population growth rates have so far resulted in the absolute numbers of Navajo-speakers remaining approximately stable.

A recent phenomenon is the development of organized concern for environmental issues such as pollution in the Aneth, Utah, oil fields, a proposed toxic-waste-disposal site near Dilkon, Arizona, and logging of the Chuska Mountains forests. A group of younger Navajos have formed Diné CARE to oppose such actions, drawing upon traditional Navajo values that show respect for the Earth as well as upon contemporary mainstream scientific and political awareness (LaDuke 1996).

Conclusion

Geographer Imre Sutton (1994, 265) wrote, "land still remains the crucial issue in linking Indians to their past, their religions, and their lifestyles." This certainly is true of the Navajo. As Peter Gold (1994, 17) put it, "Clearly, in every respect, . . . Navajos see themselves as crystallizations of the substance and energy of their places on earth. . . . With such deep physical roots comes an even deeper conceptual connection with place; their material and spiritual works all reflect this connection."

In terms of the combination of population size, geographic extent and contiguity, and homogeneity, the Navajo homeland is the most notable of America's ethnic homelands. The size of West Virginia, its extent approaches that of the Hispano homeland in the same general region (Nostrand 1992) and exceeds that of the Cajuns (Comeaux 1992). But unlike the Hispano or Cajun homelands, the Navajo is characterized by a contiguity of land tenure and by a population that is nearly 100 percent homogeneous over the largest part of its extent, diluted only by a handful of non-Navajo Bureau of Indian Affairs,

Indian Health Service, and other government employees, plus a few missionaries and commercial persons.

For the Navajo, there is an unusually strong sense of homeland, equaled, if at all in the United States, only among certain other Native American groups. This sense is fostered by three circumstances: (1) the perception of sanctity of the foreordained Dinétah, laid out specifically for the People; (2) the creation of a legal and political entity, Diné Bikeyah—the Navajo Indian Reservation (plus the satellite reserves and allotted and purchased lands) in which Navajos have essentially exclusive rights of residence and land use and, today, a considerable degree of administrative autonomy in the form of the Navajo Nation (Iverson 1981, 1983; Harvey 1996; see, also, Sutton 1991); and (3) the continuation of many aspects of a traditional culture that exists in clear contrast to that of mainstream America.

CHAPTER TWELVE

Mormondom's
Deseret Homeland

· ·

Lowell C. "Ben" Bennion

Utah as Zion: Images and Icons of Deseret

In the minds of most Americans, Utah long ago became synonymous with Mormons and Mormon Country—the heartland of the Church of Jesus Christ of Latter-day Saints (and the home of the redoubtable "Polly Gamy"). In the minds of most members of this "worldwide church," 85 percent of whom now live outside the Beehive State, Utah signifies Zion, if only as the headquarters of their religion. Almost a third of America's "saints" still reside in Utah, and a large diaspora of Beehive-born members claim it as an ancestral homeland. No other state has such a long and strong connection with a specific people bound by a common bond. Even U.S. Mormons living outside of the inter-mountain West often refer to "the Utah church" as if it represented a distinct body of saints. Where else but in Utah (or Idaho) do Americans identify each other so readily on the basis of religious affiliation and activity?

As a long-time but non-Mormon Utahn emphasizes, "The Mormon presence is the preponderant cultural, political, and economic fact of Utah. The Mormons are as indisputably a part of the structure of Utah life as the weather" (Lyon and Williams 1995, 11). No one senses the Latter-day Saint (LDS) influence on the local climate more acutely than the non-Mormon, or "gentile," minority (25 percent of 2,000,000 in 1996), which also considers Utah its home.

The name *Utah*, derived from the native Ute population, conjures up certain physical images besides the cultural icons associated with religion. Many of John Telford's photos capture the striking contrasts in topography found across the state (Smart and Telford 1995). The varied terrain reflects the convergence of three physiographic provinces: the Great Basin, the Rocky Mountains, and the Colorado Plateau. In addition, the Mojave Desert juts into Utah's southwest corner, long called "Dixie" because of its lower elevations and higher temperatures.

Of all the icons that may say Utah, the Angel Moroni atop the Salt Lake Temple and the Beehive above the former Hotel Utah probably remain the best known. Both overlook the two city squares that epitomize Utah as a promised land—an American Zion. Together, wherever they appear, the angel and the beehive symbolize the spiritual and material essence of Mormonism (Mauss 1994). Temple Square and what I call COB Square (fig. 12.1), the latter dominated by a maze of church office buildings, draw nearly twice as many tourists as the only place in Utah named Zion—the most popular of the state's five national parks (and the only one in Dixie).

Salt Lake's Central Business District (CBD), which centers on the two church squares, functions not only as the hub of Utah but as the crossroads of a much broader Mormon culture region (Meinig 1965). As Zion's premier city, little Salt Lake proper (1994 population: 175,000) exerts a powerful influence across an area much larger than Utah. The city lies at the center of the Wasatch Front metropolis, which reaches 50 miles both north and south and numbers more than 1,500,000 residents. Salt Lake, moreover, dominates the rest of Utah and directly affects all counties of the intermountain West with a sizable LDS minority.

Mormon Country today has a recognizable if rough correlation with the vast western territory that Brigham Young in 1849 asked Congress to accept as the State of Deseret for the saints who had gathered to the Great Basin (map 12.1). Outsiders may think Deseret is a misspelling of *desert*. While the territory, both then and now, embraces much desert, the name is an obscure term for "honeybee" which appears but once in the Book of Mormon. But more Salt Lake businesses carry this name than is the case with any of the area's other icons—Beehive, Ensign, Zion, Pioneer, Ute, and the like—perhaps because so many Deseret companies are owned by the LDS Church. Only two places, a village near Delta and a peak west of Tooele, bear the name Deseret. Today

Fig. 12.1. Church of Jesus Christ of Latter-day Saints Office Building, Salt Lake City, Utah, completed in 1972, dwarfs all other buildings on its square as well as the temple and Conference Center. Its height symbolizes both the rapid growth of the church and the central control of finances and programs by the General Authorities. One can almost imagine the fountain in front as a "beehive" beholden to the Angel Moroni atop

nobody refers to Utah as Deseret except when singing a hymn entitled "In Our Lovely Deseret."

The term *Mormondom* serves as a convenient referent for the entire membership of the LDS Church. Although the proportion of saints in the United States currently shrinks at the rate of about 1 percent per year, Utah and the Wasatch Front will spell Zion for most of them well into the future. In spite of recent dramatic shifts in the distribution of Latter-day Saints, Deseret seems destined to remain the global center of Mormondom and the primary homeland for American saints (Bennion and Young 1996; De Pillis 1996).

A Utah-centered Mormon Country clearly fits the criteria used in this volume to define *homeland*. First, from its start as a part of the American West, Deseret has attracted an unusual mix of peoples—both Mormon and gentile—who have shaped a regional society unlike any other in the nation. Second, as the ruling majority in Utah Territory, nineteenth-century Mormons forged a "near-nation" within an expanding America. While Washington finally compelled the polygamous saints to conform to certain national norms for statehood in 1896, Utah has maintained an identity that stands out on many national thematic maps, notably in contrast to neighboring Nevada. For example, Utah has the highest percent of native-born residents (70) among all western states, whereas Nevada shows the lowest (15) (Wright 1998, 41). Third, wherever they settled, no matter what the terrain, the saints fashioned similar kinds of cultural landscapes more reminiscent of New England or the Middle Atlantic area than those found on the Nevada side of the Great Basin. Fourth, in the settlement process Mormons acquired a fondness for their mountain-enclosed valleys that manifests itself in many facets of their life, perhaps most obviously in the hymns they sing (Kay 1995). Finally, these distinct qualities of the Deseret homeland have a tangled history that spans, as of 1997, a much celebrated sesquicentennial.

Historians usually divide this period into three half-centuries, each marked by major changes in the evolution of Utah as a Mormon-gentile Palestine. These eras provide a convenient frame for examining the dynamic aspects of Mormon Country as an American homeland.

---

the temple to the west, especially with the church's recent conversion of the intervening section of Main Street into a plaza. Photograph 50002-21 courtesy Gary B. Peterson, May 2000.

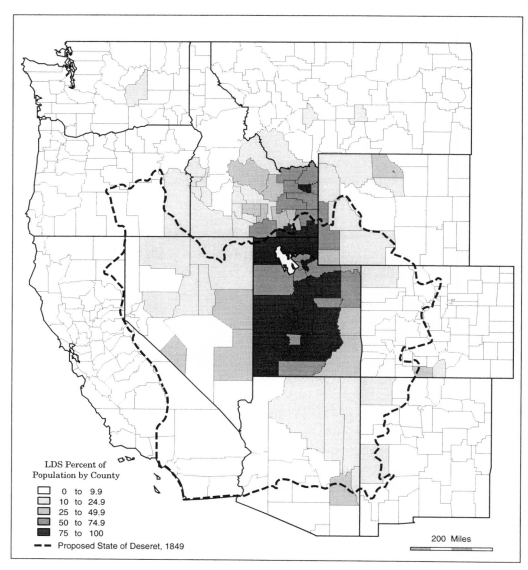

LDS Percent of
Population by County

☐ 0 to 9.9
☐ 10 to 24.9
▨ 25 to 49.9
▨ 50 to 74.9
■ 75 to 100
— — Proposed State of Deseret, 1849

200 Miles

Map 12.1. Mormondom's Deseret homeland, 1990. More than half of all American Mormons currently reside in Utah and the seven intermountain western states (including Montana) grouped around it. This map shows how much the Mormon percentage varies by county inside and outside of Utah. Wherever they comprise at least 10 percent of the population, the saints also represent the single largest denomination because so many westerners are unaffiliated with any church.

The territorial era (1847–96) established the basic shape of Deseret, but understanding its patterns requires a synopsis of the saints' futile search (1830–46) for a permanent Zion within the eastern United States. By the time of its admission to the Union, Utah had emerged as the imperial center of a fragmented "Great Basin Kingdom" (Arrington 1958) or "Rocky Mountain Empire" (Taylor 1978) that reached from Alberta, Canada, to Chihuahua, Mexico. Statehood, however, brought less independence than LDS leaders anticipated, and for the next half-century (1897–1945) Utah struggled as an economic colony of the East. Only since World War II (1946–97) has Utah emerged as an urban commonwealth more firmly in control of its economy (Alexander 1995a). Ironically, during this period Utah has become the most Republican state in the nation, the only one in which President Bill Clinton placed third (behind Ross Perot) in the 1992 election (Alexander 1995b).

Eastern Origins of Western Mormons, 1820–46

Mormonism began within the religiously "burned-over district" of Upstate New York, along a 150-mile axis that stretched from Joseph Smith's home near Palmyra to his wife Emma Hale's home in Harmony, Pennsylvania, on the Upper Susquehanna. Ten years (1820–30) elapsed between Smith's first vision and his founding of a Church of Christ. During that decade an angel named Moroni directed him to buried records purportedly written by early immigrants from ancient Israel and their descendants (known as Nephites and Lamanites). Smith translated the writings as the *Book of Mormon* (named after the same Moroni's father). The bulk of the volume spans a millennium (600 B.C.–A.D. 400) and resembles an amalgam of the Old and New Testaments set vaguely in the Western Hemisphere (Shipps 1985 and 1998; Conkin 1997).

Between his 1820 vision in a sacred grove near his home and his 1844 martyrdom in western Illinois (at age 38), Smith sought to restore a pristine version of biblical Christianity. Geographically, the "spirit of gathering" became the most important of Mormon doctrines because it mandated a homeland for the restored church. "Joseph Smith turned [American] space into a funnel that collected people from the widest possible periphery and drew them like gravity into a central point" (Bushman 1997, 5). Ironically, the gathering kept the first generation of converts in constant motion. They moved from their place of conversion in the eastern United States or western Europe to as many as five destinations during the first 15 years (1830–45).

Most accounts give the impression that the Mormons were "Driven from New York to Ohio to Missouri to Illinois to Winter Quarters on the banks of the Missouri" by "the mobbings and burnings and killings" of their enemies (Smart and Telford 1995, 192). In fact, although the infant church experienced opposition in New York, Joseph Smith never viewed the Palmyra-Harmony hearth as Zion nor intended to stay there. For him, a future homeland lay on the then western borders of the United States, close to relocated Indians (or Lamanites) whom the saints would "redeem."

Early in 1831, before declaring western Missouri *the* Zion of the New World and the site of the original Garden of Eden, Smith moved his followers to the Western Reserve of Ohio—a temporary foothold for planting the infant faith farther west. His choice of the Kirtland area resulted from early missionaries' unusual success in that region. From there he laid plans for platting a "New Jerusalem" in Missouri.

Significantly, Kirtland became a more important center than did Independence. In the former the saints built their first temple (1836). From his Ohio base, Smith directed the gathering of mostly Yankee converts to Jackson County, Missouri, the "center place" for the saints in the midst of skeptical and more numerous southerners. The two cultures inevitably clashed over fundamental geopolitical issues: "the control of territory and of the character of society therein" (Meinig 1998, 91).

The outnumbered Mormons proved no match for the Missourians, who viewed gathering and theocracy as most un-American. The saints were forced to flee north in late 1833, first into adjoining Clay County and then, 30 months later—after being asked once again to leave—into an emptier Caldwell County. There they started a new "City of Zion" known as Far West, where they were joined by most of the Ohio saints (including Smith), who finally abandoned Kirtland after falling into dire economic straits during the 1837 panic. Not surprisingly, this new influx of Yankees aroused more strife, leading to the jailing of Smith and a final expulsion from Missouri in the winter of 1838–39 (Allen and Leonard 1992).

The saints' reluctant retreat from Missouri-cum-Zion took them back across the Mississippi into Illinois, to a malaria-ridden region centered on Commerce but renamed Nauvoo (or "City Beautiful"). The "Plat of the City of Zion" that Smith drafted for Independence but never fully implemented, not even in Far West or Kirtland, he now applied to an all-but-empty Nauvoo. As in Ohio, the temple served as "Mormonism's primal architectural space, as the city was its living space. . . . Joseph's temples . . . focused sacred power at a single spot" (Bushman 1997, 16).

In western Illinois converts from the newly opened (1837) mission field of Great Britain greatly augmented the Yankee "veterans" who had survived the vexations of New York, Ohio, and/or Missouri. Not all found sanctuary in Nauvoo itself; many of the gathered saints scattered well beyond the city into outlying villages, but all eagerly awaited the endowment of spiritual power and eternal marriages promised them when they finished building their temple. The greater size and dispersion of the LDS gathering in the Nauvoo area hemmed in the non-Mormons so much that the state finally revoked the city's charter in January 1845, a year prior to completion of the temple.

Before his assassination in June 1844, Joseph Smith had sensed the need to redefine Zion because of the saints' failure to establish a gathering place in Missouri. His vision probably included South as well as North America, but plans for locating new gathering places and creating more stakes, or clusters of congregations, and temples focused on the United States. Ironically, Mormon thinking still envisions Independence as the center-stake for Christ's millennial reign in spite of its function as the headquarters of the Reorganized Latter-day Saints (RLDS), fellow believers who refused to follow the Mormons across the Missouri to Utah.

Nauvoo replaced provisional Kirtland as church headquarters and as *the* temple city, but Smith never viewed it as a permanent gathering place either. That fact made it easier for the saints to abandon Nauvoo as soon as they had constructed a temple there. Indeed, Brigham Young (Smith's successor in the eyes of the LDS but not the RLDS) saw the pending uprooting as a "glorious emergency." "Far from causing the exodus . . . , the mob . . . provided an opportunity for announcing a decision already made" (Esplin 1982, 102).

After Nauvoo lost its charter, Young launched the Great Western Measure, a plan for finding a more promising Zion in an empty, isolated area farther west. He ruled out Texas and Oregon because they already had too many "mobocrats." After poring over the reports of John C. Fremont's 1843–44 expedition, Young and the other LDS apostles decided upon a remote region of Mexico's Upper California—centered on the Great Salt Lake between the major concentrations of Ute and Shoshone Indians—as a final center for a Mormon homeland.

Thus months before leaving Nauvoo early in 1846, Young had located the general area where he thought his people could secure a territory that they would not have to share, even if—as expected—it became part of an expanding America. On Fremont's maps the LDS found "the place which God for us prepared, Far away in the West," to cite a line from their signature hymn—"Come, Come, Ye Saints." That place lay somewhere between Utah Lake and the Bear River Valley, near the same "Big Salt Lake" where another New Yorker named Smith (Jedediah) had made his "home" 20 years earlier while exploring a huge "wilderness" (Morgan 1953).

The choice of a Western home by the "Pioneers of Israel" would soon take on even more geographical irony: within a year of their arrival in Mexico's Upper California, they again resided within a greatly enlarged United States; and the major overland route to California's rich Mother Lode—where a few released members of the Mormon Batallion worked for John Sutter—passed directly through Salt Lake City, making it a bona fide crossroads rather than a remote outpost in the Far West.

The saints' relocation to "Great Salt Lake City of the Great Basin, North America" (so named within a month of their 24 July 1847 arrival) resulted from the "animosities that had been deepening for more than a decade: early rumblings in Ohio, suddenly severe in Missouri, disastrous in Illinois" (Meinig 1998, 89). That historic date, when Young designated the Salt Lake Valley as the right gathering place, marked the end of a series of painful uprootings. Pioneer Day, celebrated every July 24th since 1849, became a "Mormon Fourth of July" (Stegner 1942, 234) "for creating and maintaining the collective memory of the Latter-day Saints" as a covenant people (Olsen 1996–97, 174).

Mormons often celebrate the Pioneers' "Great Trek" across the Great Plains and Rockies as if it began in the winter of 1846 and ended in the summer of 1847. In fact, most of the 16,000–18,000 LDS who vacated the Nauvoo area arrived in Salt Lake in 1848–52 after spending at least a year in one or more of a hundred camps strung out along the Missouri River north and south of Omaha–Council Bluffs (Brown et al. 1994, 74–75). Some might have stayed longer had Young, in 1852, not ordered them to march west to the new Zion. The Mormon Trail paralleled the Oregon, but the saints stayed north of the Platte River to minimize contact with the Oregonians (many of them Missourians) and to maximize forage for their livestock.

Planting Zion in a "Great Mountain Desert," 1847–96

Before the Mormon "exodus of necessity" from Nauvoo (Bennett 1997, 360), LDS officials must have sensed that their prospective homeland resembled an inverted version of biblical Palestine (map 12.2). Whether that spatial parallel, with a freshwater sea (Utah Lake) flowing north instead of south into a dead sea (Great Salt Lake), influenced their selection of a Great Basin locale, no one has determined. But they renamed the river that linked the two seas the Western Jordan to distinguish it from the Eastern Jordan of Palestine (Madsen 1989).

About two miles east of the Jordan River, at the north end of the valley, the Pioneer Company laid out a spacious grid for a City of Zion on City Creek and centered it on Temple Square. Siting this "temple city" (Barth 1975) marked the start of an American Israel that by the end of the nineteenth century dwarfed the Palestine of the Old World.

In recasting his classic essay on the Mormon culture region, Meinig identifies nine aspects of the "Mormon System of Colonization" that transformed the habitats of Paiutes, Utes, Shoshones, and other native peoples into an unusual American homeland in less than fifty years (Meinig 1998, 96–104). My grouping of these features under four headings shows how the saints made "a genuine people" within a "great mountain desert" kingdom (Mulder 1957).

Salt Lake City as Urban Nucleus

From the start, Salt Lake served as the nerve center of a network designed by Brigham Young to create a Great Basin Kingdom within the American empire. As the main decision-making place, Salt Lake controlled the rest of the territory in every respect, often generating resentment. Even with the rise of Ogden and Provo as important secondary hubs along the Wasatch Oasis, the capital's primacy persisted. By 1890, about 50 percent of Utah's population lived in the four Salt Lake–centered counties that still define the Wasatch Front. This split has created tensions in the urban hierarchy between the cen-

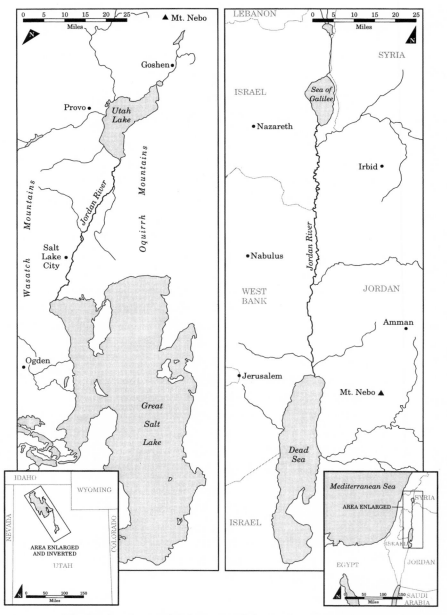

Map 12.2. The Jordan River of Palestine and the Jordan River of Utah. The two "Palestines" are mapped at the same scale and the Salt Lake Valley is inverted to facilitate comparison of the Jewish and Mormon homelands.

tral "beehive" and a host of smaller ones, and also a certain schizophrenia in individuals who "wanted the status and friendship offered by conventional [urban] community as well as the freedom and economic opportunity of the [rural] desert" (Bennion 1994, 267).

Deseret and Gateways as Geopolitical Frame

The vast size of the proposed State of Deseret reflected both the relative emptiness of the Great Basin–Colorado River region and Brother Brigham's desire to keep other peoples at bay (map 12.3). He tried to relocate the capital to a more central location, placing it 150 miles farther south in Fillmore (1851–58), but the plan came to naught because of Salt Lake's headstart and superior advantages. During the same period Young also secured three peripheral points that connected Deseret with the rest of the United States: Ft. Bridger (western Wyoming) in a strategic position on the Mormon Trail, Mormon Station (near Carson City, Nevada) as a gateway to the Bay Area, and San Bernardino as a link to Los Angeles and the Pacific. By 1857, Young decided that the two western outposts had served their purposes. A majority of their settlers returned "home" just in time to confront the army sent by President James Buchanan to assert stronger national control of Utah Territory. So determined were the saints to defend their Deseret homeland that 35,000 of them abandoned their homes in northern Utah and took refuge in settlements south of Salt Lake Valley. Had Johnston's army attacked in 1858, the LDS would have torched their homes before letting federal troops touch them.

Contiguous and Discontinuous Expansion

While the frequent and sizable reductions of Utah Territory in the 1860s give the impression of a shrinking kingdom, they belie the simultaneous expansion of Mormon colonization. As fast as converts gathered to Salt Lake City, the Great Colonizer dispersed them to smaller oases located largely along a north-south axis that eventually extended from the Upper Snake River of Idaho to the Virgin Valley of Dixie and the Little Colorado River in Arizona.

Because the Deseret environment becomes exceedingly harsh beyond the Wasatch Oasis and the frost-prone granaries of Cache (Logan) and Sanpete (Manti) valleys, church authorities often had to call and cajole people to settle

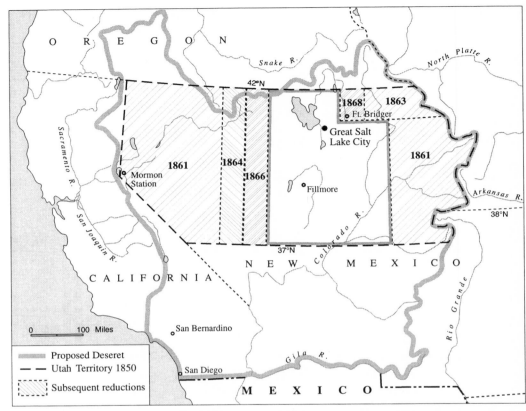

Map 12.3. The reduction of Deseret to Utah, 1849–68. Congress rejected Brigham Young's proposed state of Deseret (1849), but Young acted as though Deseret were a reality and secured peripheral outposts at Fort Bridger, Mormon Station, and a Mexican ranch on which Mormons founded San Bernardino. Following the Utah War (1857–58), Young oversaw Mormon colonization of a contracted area stretching north and especially south of Salt Lake City along both sides of the Wasatch Range. Meanwhile, Utah Territory (1850) found itself whittled away in the 1860s by a hostile federal government that authorized creation of the territories and/or states of Nevada, Colorado, and Wyoming.

these less favored lands. The Brethren promised the saints that if they exercised sufficient faith and toil, God would ameliorate the climate enough to enable them to change "Waste Places" into gardens no less verdant than the once-allegedly-barren Wasatch Oasis (Geary 1996a). As a result, "Church leaders' [later] descriptions of the Salt Lake Valley increased its aridity . . . in direct correlation to the harshness of the area they were then encouraging settlers to occupy" (Jackson 1981, 9).

The reduced boundaries of Utah never deterred Mormons from seeking footholds in other territories or states. Moreover, the federal government's antipolygamy raids of the 1880s prompted some plural families to plant colonies outside the United States, where they created miniature homelands within Alberta and Chihuahua. On the edges of Utah and beyond, LDS settlements became more scattered because gentiles had already moved in and much of Deseret was too dry and rugged even for God to intervene.

Territorial Integration and Settlement Design

The church hierarchy devised various means to achieve its grand aim of "building the heterogeneous harvest of converts arriving each fall . . . into a unified, harmonious, orderly community" (May 1977, 76). By the time of his death (1877), Young had divided Utah into twenty different stakes, regional units that approximated counties and contained anywhere from 5 to 20 wards—city neighborhoods or villages. About half of the church's 12 apostles lived outside of Salt Lake in one of the regional centers. Each stake held quarterly conferences visited by two or more of the General Authorities, and local leaders attended semiannual conferences in Salt Lake for further instruction. The church's president and presiding bishop also used both the telegraph and the railroad to integrate more tightly than wagon roads could the hierarchy of "beehives" that defined Deseret.

To promote community and unity, the church encouraged the colonizing saints to cluster in villages—invariably called towns or cities. "In Utah, Mormons for the first time constructed chapels for worship and activity, creating hundreds of little epicenters of religious life. The diffusion of church buildings necessarily flattened the religious landscape, making Mormon space more like Protestant space outside of Utah" (Bushman 1997, 25). Besides Salt Lake, only St. George, Manti, and Logan became temple towns in Utah, finishing their temples a decade or more before the capital city finally completed its own more famous one in 1893.

Although Mormons occupied most of Utah's diverse topography within a single generation, their settlements—regardless of size—looked remarkably similar. The standard plat consisted of "a gridiron aligned with the cardinal directions, of . . . large blocks (ten acres) and wide streets, subdivided into large lots (one and a quarter acres each) of alternating orientation and uniform setback of houses, with every street bordered by open ditches of flowing water" (Meinig 1998, 101).

Yet many of the smaller and even some of the larger settlements deviated from the standard pattern of Zion. Lowry Nelson (1952), a rural sociologist who helped foster the common image of compact towns, never used his home village of Ferron as a case study because about a third of its residents lived outside town on scattered homesteads. The same pattern prevailed in the rest of Emery County and in a number of other areas—notably southeastern Idaho—where homestead laws facilitated farming in a "scattered condition" (Geary 1996b). Even in Salt Lake Valley, many "country" wards were strung out for miles along a given stream or street with no visible node other than the ward meetinghouse. Despite the many exceptions to the general rule of nucleation, it does seem surprising that more dispersion did not occur after Utah's adoption of national homestead laws in 1869 (Bennion 1991; Sauder 1996).

Whether clustered or scattered, Mormon towns often split into factions based on the diverse origins of their inhabitants. For instance, "Brigham Young called settlers to Kanab in 1870 under the express condition that their new community be characterized by 'cooperation in all things'" (May 1977, 88). When Young later tried to introduce a communitarian United Order there (and throughout Utah), the town became deeply divided and remained so for years. At the same time, after the U.S. Army erected a fort near Beaver, that town's saints disagreed about how to deal with the influx of soldiers and miners—whether to fraternize or shun them. The issue merely compounded the town's existing factions. In general, it took a generation or two for the saints to achieve the unity and uniformity usually ascribed to them.

One outsider, a perceptive British traveler, recognized that Utah's population, even then, consisted of "two peoples," who "do not mingle any more than oil and water" (Robinson 1883, 140). Nowhere were "Sinners and Saints" more antagonistic toward each other than in the divided capital city and in the railroad hub of Ogden farther north. Such hostility, however, did not keep the LDS from developing a sense of home; indeed, it undoubtedly reinforced their attachment to Utah as a "mountain retreat."

A sense of homeland, of course, involved more than years of living in the same place. Even in their Great Basin Kingdom, "The Saints remained in constant flux; the continuous intermixing served to break down the walls of local insularity after Zion began to assume the character of a settled place, and made the Saint pretty much a standard product" (Morgan 1949, 191).

In fact, some pioneers moved so often they never became really rooted. Polygamist Lewis Barney, for example, changed residence 17 times between 1852 and 1894, without any call from the Brethren. He wanted "a place where [he] could have [his] family all located around [him]" but failed to find one. He lamented that his family was "scattered all over the country" (Barney 1978, 112). Despite their mobility, the busy Mormon bees built a genuine homeland "with a unity, homogeneity, order, and self-consciousness not . . . found in any other region in the United States" (Meinig 1998, 3: 104).

Making an American State out of a Mormon Deseret, 1897–1945

The apparent abolition of polygamy and church control of politics in the 1890s should not imply that Utah rapidly emerged from its territorial era in harmony with mainstream America. Most studies indicate that the transition from a wayward agrarian territory to a modern urban-industrial state required a full half-century.

Utah's rising rural population moved from farms or villages into cities during the early decades of the twentieth century. Salt Lake County's share of Utah's population jumped from about 25 to 40 percent, most of it concentrated in Salt Lake City. This increase made Utah's capital by far the largest city between Denver and Oakland. The announcer of the Mormon Tabernacle Choir's weekly radio broadcasts, which began in 1929, exaggerated only slightly when he referred to Temple Square as "the Crossroads of the West."

Salt Lake's exceptional size in 1940 stemmed from its functions as Salt Lake County's seat, Utah's capital, and Mormonism's headquarters; and from its central position along the hundred-mile Wasatch Oasis that supported Utah's richest farms, mines, and smelters. The city's wealth attracted a sizable and diverse population from outside Deseret, so that—to add another irony—gentiles have long outnumbered saints at the very center of Zion (Sillitoe 1996).

As "a gentile in the New Jerusalem," Stegner characterized his hometown as "a divided concept, a complex idea. To the devout it is more than a place; it is a way of life. . . . In this sense Salt Lake City is forever foreign to me, as to any non-Mormon" (Stegner 1969, 159). But as the city's numbers increased, so did the divisions among both "saints" and "sinners." And the differences, coupled with the Mormons' minority position in their own capital, made it easier for both peoples to bridge the deep religious divide of the prestatehood years and to cooperate in sometimes surprising ways.

For instance, in spite of their advocacy of abstinence from alcoholic beverages, the LDS hierarchy delayed Utah's endorsement of prohibition for a full decade. They gave higher priority to persuading gentile businessmen to abandon their anti-Mormon American Party and support the Republican Party. Instead of voting for state prohibition, Church President Joseph F. Smith and Apostle/Senator Reed Smoot endorsed a local option bill which, when passed, divided the state into "dry" and "wet" areas. (In Salt Lake City, "wet" votes won by a 2:1.3 margin that kept 141 saloons open [Thompson 1983].)

In addition, Mormon businessmen, acting on their own or for the church, often collaborated with Eastern corporations wanting to invest in Utah resources. Such cooperation brought about "the necessary fraud," or tampering with land laws, devised to tap the rich coal reserves in Carbon and Emery counties in east central Utah (Taniguchi 1996). Pacts of this kind placed Salt Lake in the economic clutches of the East in much the same way that the rural "other Utahs" found themselves in the grasp of the capital city.

Salt Lake never had any real rivals. As of 1940, only three other towns—Ogden, Provo, and Logan—had more than 10,000 inhabitants, and even when combined their populations equaled only half the capital's 150,000. None of the three older Utah temple towns could hold a candle to *the* crossroads city.

As Salt Lake mushroomed, the rest of Deseret remained largely rural and, in the case of the Colorado Plateau and Dixie, remote. As the urban-rural rift widened, regional divisions within Salt Lake County also loomed large. The completion of the stately City and County Building in 1894, which housed the state government until it occupied its own imposing Capitol in 1915, belied any unity implied by its name. During this period the "city," confined to the northern end of Salt Lake Valley, contained 70–75 percent of the county's population and naturally dominated the mainly unincorporated areas—referred to as "county." In addition to this north-south, urban-rural schism, tensions also developed between the east and west sides of the valley. The eastern half always had more water, better soils, and a larger and wealthier population.

The presence of a gentile majority in the city and increasing Mormon involvement in a national capitalist economy greatly altered Salt Lake's townscape. The core of the city resembled a T-square after Utah's copper king, Samuel Newhouse, developed an Exchange Place complex several blocks south of Temple and COB squares. His luxury Newhouse Hotel, close to the Federal Building, rivaled the church's new Hotel Utah—capped by a beehive embellished by American shields and eagles. Mining money also built many of the mansions that lined east South Temple, the street that linked downtown with a slowly expanding University of Utah (formerly Deseret), located below the U.S. Army's Fort Douglas. Catholics, Protestants, Jews, and Masons all added their own structures to the Salt Lake skyline to make it less Mormon.

An Expanding Zion and a Dynamic Deseret, 1946–97

Mormonism has experienced phenomenal change since World War II. Between 1946 and 1971, the membership tripled in size from about one to three million; since then it has tripled again. In the process the church has organized a ward in virtually every county of the American West and at least one stake in every state of the East. While spreading out regionally and nationally, the religion has expanded even more at the international level by establishing a presence in every country (160 at latest count) that will grant it entry. Wherever members are numerous enough to justify the cost, the church has built temples, and sometimes missionary training centers, as growth poles for a global network of "mini-Zions" designed to diffuse Mormonism (Shipps 1994).

The church's expansion has overshadowed but also initiated important changes within the LDS homeland. Meinig's breakdown of the Mormon culture region into *core, domain,* and *sphere* recognizes major differences inside Deseret by combining the two standard types of regions used by geographers

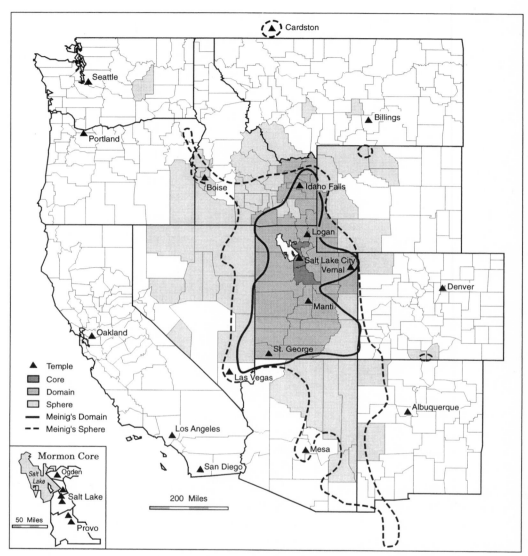

Map 12.4. The Mormon culture region, 1990. Meinig's core, where Mormons predominate but non-Mormons form a sizable and an influential minority, is the interactive node of a functional region that extends across all of the contiguous shaded counties. His domain and sphere represent gradations of 50–90 and 10–49 percent LDS, respectively. The correlation between Meinig's domain, drawn in 1965, and counties outside the core that were at least 50 percent Mormon in 1990 is striking. The sphere has undergone the most change, expanding in every direction except the southeast.

(map 12.4): the uniform type refers to any area that has one or more common traits such as a Mormon majority; the functional type defines an area tributary to a central place such as Salt Lake City's hinterland.

The core, signifying extreme concentration of people and power, comprises the same four counties usually equated with the Wasatch Front. On the map it takes the shape of a cowboy boot pointed east. This region has almost quadrupled its population since 1946 and now draws many commuters from the mountain and desert counties that adjoin it. Much of this "back country" around the core contains countless second homes built by urban "cowboys" who live and work along the burgeoning Wasatch corridor.

Within the core, Provo-Orem overtook Ogden as Utah's second largest city in the 1960s, owing to the growth of Brigham Young University (BYU, the "Y") as the church's leading university and the flourishing of high-tech firms in Utah County. The county's "Happy Valley" nickname presumably reflects the BYU influence that has made it the most Mormon (90%) county of the core. It boasts an exceptionally high level of church activity even among Utah Mormons and the state's highest birth rate.

BYU currently draws three-fourths of its 30,000 students from non-Utah Mormondom. For most of these out-of-state youth, Utah must mean first and foremost BYU and, perhaps as often as not, the state where they will meet their mate and maybe even make their home. Many Utahns, including long-time U.S. Senator Orrin Hatch, a native of Pennsylvania, can recount such a scenario (Hoskins 1996, 108–9). A Pennsylvanian of my acquaintance but, unlike Hatch, a convert to the church, denies having ever had any "pine for Zion" feeling. Yet he and his wife sent their five children to the "Y," and three of them now live in Utah with their own families.

Every August BYU hosts a Church Education Week that draws about 30,000 visitors, and during the school year faculty members give Know Your Religion lectures in LDS stakes throughout the Western states. As the flagship of the Church Education System, BYU's influence among Mormons now rivals Salt Lake's in important ways. Moreover, the church has placed its only training center for American and Canadian missionaries in Provo. Thus, well over half of the nearly 60,000 now serving missions received two to eight weeks of training at BYU before going out to proselytize.

The University of Utah, located in Salt Lake City, has become BYU's keenest athletic rival. This U vs. Y rivalry, which divides Salt Lakers and even some LDS families, reflects Utah's long-standing church-state division and two related facts: Salt Lake City still has a gentile majority, and the U's administration and faculty have become mostly non-Mormon since 1945.

Salt Lake County itself has become more diverse, perhaps even more divided, during the past five decades. Only 25 percent of the county's population now lives in the city compared to 70 percent as late as 1950. Unhappy with county government, more and more "country" suburbs have incorporated since 1980. Except along the east central bench, the county's old towns, spread out from the start, have become sprawling cities that, in one case (West Valley), can claim as many residents as Provo-Orem or Ogden. During the recent suburban surge to incorporate, the county finally moved all of its offices from the old City and County Building to a complex of new ones on the city's border with South Salt Lake. The city, it seems, has seldom sought to expand at the expense of its Siamese county (Sillitoe 1996).

With the razing and rebuilding of so much of the downtown area, Salt Lake's townscape no longer shows any sharp gentile-Mormon schism. To be sure, the church still owns and controls much of the northern part of the Central Business District. But even that end of town has undergone a radical facelift since the church erected the city's tallest skyscraper (28 stories) in 1972 to house its growing bureaucracy. The difference in size between this new high-rise and the old five-story Church Administration Building, both on COB Square, clearly mirrors the church's postwar growth.

The LDS hub has expanded outward as well as upward. Having outgrown Temple Square's historic Tabernacle, site of its semiannual conferences, the church has constructed a new Conference Center four times as large on the block to the north. West of Temple Square, it already has in place the Family History Library and Museum of Church History and Art. Still farther west stands Broadcasting House, from which Bonneville International operates a far-flung network of radio stations.

Change has become rampant lately. The present "cratered" look of the downtown area (and other county areas besides Bingham Canyon) will likely persist until the state has revamped its transportation system and otherwise prepared itself to host the Winter Olympics in 2002. Perhaps a sense of the past in the next millennium will manifest itself mainly in museums and parks constructed by the church or city as reminders of the pioneers' legacies. City Creek (just before it goes underground), Ensign Peak above the Capitol, and

Old Deseret Village near This Is The Place Monument stand out among several sites recently developed for this purpose.

Salt Lake's population, even its Mormon minority, has become more cosmopolitan despite the city's slight decline in size from 1960 to 1990. Most of Utah's non-English-speaking LDS wards and branches lie inside the city. The church organized them for members who have moved to Deseret in recent decades from Europe, East Asia, Oceania (notably Tonga), and Latin America (Brown et al. 1994, 146–47), ignoring the church's century-long "ban" on gathering to the central Zion. As recently as December 1, 1999, the First Presidency issued a directive "to reiterate the long-standing counsel to members of the Church to remain in their homelands rather than immigrate to the United States."

Beyond the Wasatch Front metropolis lies the domain of Deseret, a region that remains largely LDS (at least 50%) but decidedly rural, if increasingly nonfarm. Its two largest cities—Logan and St. George—count about 30,000 citizens each. Except for my use of county data and boundaries, the domain delimited on map 12.4 closely matches Meinig's 1965 delineation. Three counties—one each in Wyoming, Idaho, and Nevada—fall just short of the 50 percent cutoff, whereas the two Utah counties on the far side of the Colorado River record under 40 percent. A gentile's observation may explain this anomaly: "There was no way for the Church to control cowpunchers [and miners] as it controlled villagers. As a result, Mormonism was very early . . . diluted at its edges" (Stegner 1942, 282). He might have added that LDS elders had little luck in converting San Juan County's large Indian population. Artists, bikers, hikers, and river-runners now outnumber cowboys and miners, if not Navajos, in the Grand–San Juan area, but few of them seem inclined to become Mormon.

During the past 50 years most counties in Deseret's domain have struggled just to maintain their low population levels. While 7 of 29 have failed to do so, a few have doubled in size—led by Washington County, where St. George has realized its dream of becoming Utah's "Palm Springs." Many Mormon villages and gentile towns have become virtual ghosts, but others have experienced considerable change with growth and modernization. Their rural character has attracted urban refugees and retirees, from both the Wasatch Front and the West Coast, who often try to citify their previously unkempt properties with urban designs.

If Mormons form a majority within Zion's "walls"—that is, the urban core and rural domain—they thin out only gradually across the rest of Deseret. Wherever they make up 10–49 percent of the population, they constitute a conspicuous minority (map 12.4). This sphere of LDS influence now reaches no farther past the Colorado River than it did thirty years ago. But in all other directions it has made major gains, penetrating most of western Wyoming, eastern Nevada, southwestern Idaho, and even spreading into Montana and the Columbia Basin. New temples in Las Vegas, Boise, Billings, and Albuquerque herald a striking extension of Zion's "curtains" since 1965. The last two fall in the small size category started by the current LDS president, Gordon B. Hinckley, soon after he assumed the mantle of prophet (and "amateur architect") in 1995. His "temple-building blitz" (Ostling and Ostling 1999, 148) is the most visible sign of the church's determination to disperse Zion across North America and the rest of the world (Hinckley's keen sense of place prompted him to dedicate LDS temple number 100 in Boston, Massachusetts, within the New England homeland of Joseph Smith and Brigham Young).

Modern Deseret as Mormon Sanctuary

The celebration of Utah's 1996 Statehood Centennial featured *Faces of Utah*, a profile based on more than half a million responses to an invitation to write about "what it means to me to be a Utahn" (Hoskins 1996, 9). While waiting for its publication, Rena Christensen and I conducted a similar but much smaller survey. We wanted to know if transplanted Utahns like ourselves (in California) felt the same or differently about the state as lifelong or current residents. Even from an informal canvass of a few score friends, we found it difficult to generalize about where U.S. Mormons choose to live relative to the Deseret homeland.

The experience of just one family illustrates the difficulty. An LDS couple began and recently ended their married life in western Montana, barely within the Mormon sphere. They raised their ten children for the most part in Salt Lake Valley. Half now reside inside and half outside of Deseret. Moreover, five of them remain active members, while the rest have drifted away from the church. No correlation exists between their present place of residence and degree of commitment to the LDS faith, based on their response to our questionnaire (the source of subsequent quotations unless otherwise noted).

Deseret's numerous expatriates often express mixed feelings about Utah, depending in part upon why they left the region in the first place. If they married and moved away to escape a provincial family or neighborhood, they may miss the mountains but not enough to leave a socially more diverse and tolerant place on the West or East Coast. If they left primarily for employment opportunities, they may return—like homing pigeons—the month they retire, even if it means leaving a sunny Livermore in California for a frosty Logan in Utah. Of course, cashing in a high-priced suburban home in California or New York for a less expensive one in Utah may provide enough equity to compensate for the marked change in climate.

Other "exiles" from Zion have realized that with Utah's booming economy, since the mid-1980s, they can go home again to work. At the same time, they can act upon their strong attachment to place, "where history, memory, myth, and landscape all mesh in a complex web of emotion." While they may regret the decline of the traditional rural landscape in their absence, "The loss of a portion of the past is a small price to pay to be home" (Hafen, quoted in Lyon and Williams 1995, 970, 974–75). Judging by Utah obituaries, some transplanted Mormons may not return to their part of Zion until shortly after their death. But even this posthumous act implies a preference for a mountain-valley home.

One of our respondents reported: "As far back as I can recall, Utah seemed like home to me, or as much like home as any place could be to a person who moved often. When our family [finally] moved to Utah when I was sixteen and a half, I had a distinct feeling of coming home." His Mormon parents had grown up outside Utah but had met and married in Salt Lake and retained ties to the same Zion where he has now rooted his own family.

The yearning to return to Deseret must be one reason why other American saints often refer to "Utah Mormons" as a breed apart. Transplanted Utahns frequently dominate ward and stake leadership positions outside of the Mormon culture region, especially in the eastern states, where the church remains comparatively weak. This perhaps unconscious dominance sometimes creates tensions between them and relatively new members. The tendency to pine for Zion may stigmatize many of the diaspora as sojourners in the eyes of those saints, often converts, who are rooted outside of Deseret.

"Ethnic Mormons" can be defined as any LDS who have long and strong attachments to Deseret. They may not relate to the region as a whole as much as they do to a certain locale within the core, domain, or sphere—that is, wherever they feel most at home. They undoubtedly share naturalist Terry Tempest Williams's view (quoted by Webster 1996, 57) that Deseret or "Utah . . . always has been [the promised land]. . . . Brigham Young said, 'This is the place,' and there are those of us who still believe that. [My identity is] more than history or religious affiliation, or even this landscape. It's all these: Family, Religion, Place. They can't be divided."

Another Tempest, a contractor who lives in Salt Lake County and likes to hunt, "can drive for 30 minutes and shoot in any direction." Whether nature writer or contractor/hunter, each Tempest has the "centered" feeling shared by most Utahns but eloquently expressed by a native Dutchman and emeritus English professor: "I feel about Utah what Hawthorne felt about New England: It was, he said, 'as much of the United States as my heart can hold.' During seventy years of residence here I have sent down deep roots which have flowered in rich experiences embracing family, community, and collegial associations in a beautiful natural setting with a strong sense of the past. During several long absences . . . Salt Lake City has been the unmoved leg of the compass as the other leg scribed a wide circle. I have always felt *centered* here."

Many members of the gentile minority also feel "centered" in Utah but often for other reasons than those of the Mormon majority. Some gentiles have developed a love-hate relationship with the state, finding "the dominant culture close-minded and cruel." But many others voice pleasure in having "the chance to live in beauty, among loving, caring people, in greater security and with less stress" (Hoskins 1996, 116, 183). The one aspect of Utah that all gentiles seem to like is its stunning natural beauty, and many of them have become involved in "the battle for Utah's wilderness" or "soul" (Hoskins 1996, 134).

As an astute observer noted 40 years ago, "the nation might reincorporate Mormondom into itself, but it could not break the primary association of the intermountain region with Mormon self-consciousness" (O'Dea 1957, 114). That bond enabled the saints to evolve an identity that has persisted in spite of the constant influx of gentiles into Utah, the dispersion of Deseret Mormons across the nation, and the global expansion of the church. If Utah's LDS have become less peculiar and more American in the process, their attachment to an intermountain Zion has not waned.

The boundaries of the Mormon culture region now approximate those of the Deseret area claimed by Brigham Young 150 years ago (maps 12.1 and

12.4). He might well accept the Snake River Valley as compensation for the apparent "loss" of the southern ends of Arizona and California, where absolute numbers of LDS—but not percentages—are quite high. The steady expansion of the Mormon sphere and the rising density of the core, if not the domain, have made Deseret stronger than ever. The continuing concentration of church power in Salt Lake has made the city an international crossroads for more than 11 million Latter-day Saints.

Mormonism's many ironies have combined to create a homeland of paradoxes. After more than 150 years of occupying the valleys and mountains where basin, plateau, and Rockies converge, Mormons have created a human homeland as complex as Utah's physiography. For all their international connections and experiences, they impress many outsiders—even non-Utah members—as somewhat smug and provincial. The Mormon majority that created this American homeland still shows defensive signs of discomfort with the gentile minority in its midst. The church even has difficulty reaching accommodation with its own "fundamentalists" and "intellectuals"—both viewed as suspect or apostate. While welcoming the Winter Olympics in 2002 as a great opportunity to display its "Ensign" to the world, Zion remains wary of "Babylon" and its worldly influences.

If and when Christ returns to earth to embrace his saints, he may have to head for Utah instead of Missouri, given Deseret Mormons' fondness for their mountains. As one LDS poet exclaimed, "Utah mountains, in some crooked nook of time, travel with me, they and their holdings, scribbled over anywhere else I land. My own stereopticon imaginings let me always be at home" (Thayne 1998, 254). But her view holds true only if she doesn't have to stay away too long, since, "in important ways, good [or ethnic] Mormons never leave home" (Meinig 1997, 2). Many of them have great difficulty in leaving their Deseret homeland for longer than, say, two years without at least a temporary return to Zion. As two well-known religion writers observed, "for members in Utah [and surrounding states], Mormonism is not just a shared creed but an ethnic identity and a family heritage, with a good dose of frontier nostalgia mixed in" (Ostling and Ostling 1999, 379).

CHAPTER THIRTEEN

California's Emerging Russian Homeland

. . . .

Susan W. Hardwick

Russians in California's Central Valley are a distinct group of people with a common past, a unifying folk culture, and a well-established religious belief system. They are defined in this analysis as Slavic Russians, not Jewish Russians, who identify themselves more often as "Jews" than "Russians." Most Slavic Russians belong to the traditional Russian Orthodox Church or have converted to Protestantism, usually the Baptist or Pentecostal churches. While Russian Jews have settled primarily in California's largest urban centers in the San Francisco Bay area, Los Angeles, and San Diego, ethnic Slavic Russians have become prominent players in the formation of the emerging Russian homeland expressed in the landscape of smaller cities located in California's Central Valley (map 13.1).

Russians in California's Central Valley currently reside in a unified place, a contiguous and well-bounded region that stretches north and south of the primary node of Russian cultural and economic activity located in Sacramento. Despite their recent arrival in the United States, Russians in California have already bonded with place in their adopted homeland. The dream of "coming to America" filled the hearts and minds of the vast majority of these new arrivals on the American scene for most of their lives in their native land, no matter where they lived. The often acute political, religious, and economic challenges of their daily lives in the former Soviet Union forced many to leave in search of a safe and sane new life far from the persecution of their religious belief systems.

This unique set of push factors created a little-known yet vibrant incipient Russian homeland in California. The Russian sense of place and emerging cultural landscape in this new homeland has developed much more rapidly than that of most other immigrant groups. Of particular note is the deep Russian attachment to the Sacramento and San Joaquin rivers, underscoring a strong bond with place in the Central Valley (fig. 13.1).

Despite sometimes stereotypically Slavic images of urban landscapes in neighborhoods such as San Francisco's Russian Hill, New York City's Brighton Beach, and West Hollywood, central California has become the most frequent destination for non-Jewish Russian refugees in the post-Soviet era. Indeed, in 1990, at least 60 percent of all Russian refugee destinations in the country were to California, followed by New York (20%), Massachusetts (6%), and Illinois (4%). Russians are among the oldest of all refugees in California, with a median age of 31.4 years for women and 30.1 for men. Eight percent are over the age of 65. Most are skilled blue-collar workers with large families (California Department of Refugee Services 1991). Exact counts of refugee arrivals in California and elsewhere in the United States, however, do not reflect the total number of people now living in Central Valley cities because many Russians arrived after 1990. The problem of quantitative documentation of Russians in various parts of the country is compounded by their relocation patterns after arriving in the United States. Many rapidly become secondary migrants, moving to Central Valley cities after their initial settlement in San Francisco or Los Angeles. It is estimated by refugee resettlement agencies and the Immigration and Naturalization Service Office in Sacramento that between 45,000–50,000 people of Russian ancestry resided in the valley in 1998.

The Russian experience in California also illustrates control of place. Although not yet involved in local or regional politics, Russian immigrant "control" is expressed by a strongly felt and unifying set of attitudes about territoriality particularly through property ownership. Purchasing small houses in low-income suburban neighborhoods in such places as Sacramento, Fresno, and Bakersfield has been a primary goal of Russian immigrants since their earliest arrival in the region in 1912. A profound belief in the value of owning land is no doubt a response to their inability to buy property in the former So-

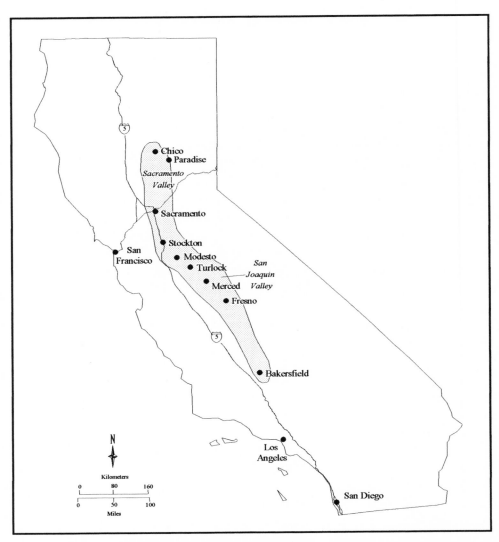

Map 13.1. California's emerging Russian homeland, showing settlement nodes in the Central Valley, 1998.

Fig. 13.1. California's Sacramento River. Most Russian immigrants in California's Central Valley who came indirectly from Siberia or central China never saw the Volga River, yet virtually all knew songs and stories about the Volga. They transferred feelings about the sacred Volga to California's Sacramento River, the chief symbol of attachment for ethnic Russian Americans in their emerging homeland. Photograph of the Sacramento River in 2000 provided by Susan W. Hardwick.

viet Union and their desire to establish roots in a new land they never intend to leave.

Time is especially fascinating as a component of the Russian case study. Although it could be argued that the strongest bonds between people and place require centuries to develop (as in the case of Hispanos in New Mexico and Cajuns in Louisiana), the Russian experience offers a unique model of rapidity and intensity of homeland development and evolution. Because their departure from the former Soviet Union often was abrupt, unexpected, and dramatic, and because of lifelong, positive attitudes about "the American dream," this immigrant group has rapidly formed a strong attachment to its new homeland. Russians bond with their new environment, it is thought, much more quickly than other groups. Indeed, in America's often chaotic multicultural society, the rapid formation of bonds with new places may become the norm rather than the exception. The emerging Russian homeland in central California may provide a model for things to come.

The Early Russian Nucleus in Central California

Unlike other immigrant groups that came to the Atlantic seaboard of the United States, Russians first settled on the Pacific coast. They established their earliest California connection in 1812 on a windy plateau north of San Francisco Bay (Essig 1933). They eventually sold this settlement called Ross to early pioneer and visionary John Sutter in 1841 (Gibson 1976). Sutter hauled Ross's building materials, animals, food products, even its redwood furniture overland to his newly developing town of Sacramento. Little did he know that over one hundred years later California's future capital city would become a major North American settlement node for Russian immigrants (Hardwick 1993).

As early as 1828 Russian explorers visited the Sacramento Valley on a hunting expedition with the Hudson Bay Company. While in the area they named the American River *Ojetska*, meaning "land of the hunter." In fact, the earliest Spanish map drawn of the area labeled the American River "Rio Ojetska" (Kantor 1880). No large-scale settlement of Russians occurred in California, though, until after the turn of the twentieth century because only a small number of Russians remained in the state after the sale of Ross.

An investigation of naturalization records, census information, newspaper accounts, church records, and other archival materials revealed that Russians first established a settlement nucleus in central California after 1912 (Larkey and Walters 1987). Interviews with long-term residents of the West Sacramento suburb of Bryte confirmed this date. Some newly arriving immigrants bought sites in the partially drained land along the flood-prone Sacramento River before it was fully dry and ready for settlement. After reclamation, Russians constructed new homes on land just west of the river. Apparently the early Russians had been encouraged by a local real estate company to buy land "sight unseen." Letters written in Russian and delivered by hand to Slavic neighborhoods in San Francisco and Arcata are still in the historical files of the West Sacramento Land Company. Other "desirable" white immigrant groups such as Italians and Portuguese were encouraged to settle permanently in the area. Land was affordable, good hunting and fishing were accessible, and jobs were plentiful at the Southern Pacific Railroad yard just across the river in nearby downtown Sacramento.

The earliest Russian settlers in Bryte were Russian Orthodox. They had opposed the Russian Revolution that began in 1917. To escape political persecution by the new Russian Communist government, most relocated in Manchuria and central China before coming to the United States. Some arrived by way of Vladivostock, Japan, and Mexico. A large number of these Russian immigrants had originally been employed in the Hawaiian sugarcane fields and eventually found their way to San Francisco (Nordyke 1977). One resident of West Sacramento who arrived in California from Hawaii recalled: "We left my home on a steamer called the 'Maru' and traveled to Hawaii. On the ship, we Russians had full freedom but the Japanese and Filipino workers had to stay below deck. What an exhausting trip it was, and my sister and I were so seasick. We really did hard work in the [sugarcane] fields, you know, and we stayed in Honolulu altogether from 1910 to the end of 1911 and then we settled in San Francisco with my father. We bought this house in Bryte in 1920" (Domasky 1983).

In the mid-1920s, this early group of Russians in the Sacramento Valley constructed the first Russian church in the region. Designed by a Russian high school student in his school shop class, this Russian Orthodox Church was built almost entirely by the women of the community using local materials (Kondratieff 1976). When the church was consecrated in 1926, the small group of believers gathered to celebrate. They planted cypress tree hedges and flowers and built fences. Every Sunday Russians living in West Sacramento today conduct lengthy services in the Old Church Slavonic language much as the original settlers did. As in the early days, there are no benches for worshippers. For over three hours men stand on the right side of the church and women on the left. Dinners are still served in the meeting room just as they were more than 75 years ago.

Russian immigrants continued to trickle into the Sacramento area in the 1920s and 1930s. After World War II, the last large Soviet-era group arrived, changing the community drastically. Because of their anti-Communist beliefs, these families were also forced to leave Russia before and during the revolution. They were Baptist Russians who had lived for many years in the Sinjiang Province of China. One of their leaders in the post–World War II years, Mikhail Lokteff, remembers the dramatic story of his family's resettlement in

California: "It was very hard for my family to leave Russia after the revolution. But my grandfather had been imprisoned and killed by the Communists and they just had to get away. My parents traveled into China and finally settled in the province of Sinjiang in the tiny village of Inning. This is the place where I was born. My group arrived here in Bryte on Christmas Day in 1950" (Lokteff 1984).

Lokteff's memories underscore the trying conditions that many Russians living in California today escaped. Their shared experiences forged religious and family bonds that remain largely in place in California today. Many Russian immigrants experienced persecution for their religious and political beliefs before arriving in central California. Most came to the United States following Soviet President Mikhail Gorbachev's startling announcement in 1987 that victims of religious persecution could leave the Soviet Union—for the first time since the Revolution ended in the early 1920s. Russian Baptists who relocated to California in the postwar years constructed in the 1950s a small brick Russian Baptist Church directly across the street from the original Russian Orthodox Church in West Sacramento. Early baptism services were held on the banks of the Sacramento River. The young Russian couple shown in fig. 13.2 exemplifies the ongoing importance of religious rituals and ceremonies.

Through inertia West Sacramento has become the preferred destination for the majority of Russian Christians arriving in the United States in the post-Soviet years. After initial arrival in California, many choose this "community of congregation" (Vance 1976) as their permanent home. Others move to Sacramento, begin their adjustment to American life, and then move on to neighborhoods in Fresno, Bakersfield, Chico, Marysville, and other smaller cities in the Central Valley.

Ensuring the continuing importance of Russians in California's Central Valley cities is the ongoing chain migration of new Russian arrivals who have been encouraged to immigrate by the United States government. Refugee policies have strongly favored family reunification as a top priority since the late 1980s. These policies, family and religious connections, and California's perceived climatic and economic amenities continue to bring new Russian residents to Sacramento and elsewhere in the Central Valley. Fresh from the "old country," these new arrivals contributed cultural and religious revitalization to the emerging Russian homeland in the late 1990s.

Fig. 13.2. Russian-American couple in Fresno, California. Russian cultural traditions and religious practices are expressed in central California much as they were in Russia and the former Soviet Union. This Fresno couple represents the enduring cultural traditions shaping Russian life and landscape in the region. Photograph by Nadia Bolyshkanov, 1989.

Religious connections are especially important in explaining the continuing attraction of Russian immigrants to central California. For the past ten years, area churches have sponsored Russian refugees and their families and have provided assistance with the complex resettlement effort. Regional offices of two Christian agencies, World Relief and the Lutheran Social Services, have been particularly helpful in sponsoring follow-up support services for new Russians arriving in the Central Valley.

Predating and complementing the work of local churches and international relief organizations has been the work of Word to Russia, a religious outreach effort housed in Sacramento. Since 1972 this group has regularly disseminated Russian language music and messages to remote places in the former Soviet Union. Despite the political and social changes in the Russian federation in the post–Cold War years, the staff at Word to Russia's recording studio and publishing offices in West Sacramento still record and broadcast taped Christian programs to Russia. Word to Russia also sends well-trained bilingual "reverse missionaries," immigrants who fled the Soviet Union, to help spread the word about their religious faith to remote areas in Siberia, central Asia, and the Russian Far East.

Numerous people living in the former Soviet Union who first heard these broadcasts were initially drawn to a new life in Sacramento. Others were told about these programs and about the Russian-American founder, Mikhail Lokteff, in American refugee camps in Vienna and Rome. According to one recent Baptist immigrant now living in West Sacramento: "It was so exciting to finally be in this place. Sacramento is like the 'promised land' to us, you know? We all felt like we already knew Mikhail Lokteff and wanted to meet him and thank him for everything he has done for us during our difficult years in Russia" (Morgunov 1990).

Russian Pentecostal radio programs have also reached deep inside the former Soviet Union. The *Voice of Truth*, also beamed from Sacramento, has been broadcast to Russian listeners for more than 40 years. These long-term, spiritually based linkages with California have helped new immigrants feel a part of their new home space quite rapidly. Many of the new arrivals, in fact, consider the Central Valley as their new "Mother Russia." Religious connections forge strong ties among immigrants and their families, and emotional bonds within religious groups intensify as individuals struggle to establish new lives in an often challenging new place.

Among the post-Soviet newcomers and older immigrants, shared hardships in Russia and China forge a tight bond between families and religions in the region extending south from Chico and Paradise through Sacramento to Fresno and Bakersfield. And contrary to what one might expect, religious differences between various Russian Christian groups have tightened ethnic bonds rather than weakened them because they allow each group to cling, almost defiantly, to their own religious identity.

Today, members of the Russian Orthodox and Baptist congregations strongly and loudly maintain their Russianness at all costs. A story told often at Russian dinner parties in Sacramento (and in Siberian cities and towns) concerns a Russian man discovered on an abandoned desert island. When asked by his rescuers why he had built *two* churches on the island when he was the sole resident, he replied: "Why that is quite simple. I built one church for me to attend and one church for me *not* to attend" (Alexseov 1985). The well-documented Russian need to resist as well as to belong has strengthened the ethnic identity of both groups. In his analysis of Italians in Boston, Gans (1962) found that divisions between religious groups intensified group solidarity. Just as Italian Catholics scorn a small group of Italian Protestant "holy rollers" in Boston, Russian Orthodox believers make fun of Baptist "jumpers" in Central Valley Russian settlement nodes.

California Place, Russian Space

While cultural, social, and economic pressures act against the preservation of Russian culture in many other parts of the United States, religion acts as an integrating force in central California. Throughout the often difficult and complex migration journey from Russia to California, spiritual beliefs helped bind and preserve immigrants and their cultural systems. Church membership implies both behavioral and social consensus of its members and conscious and unconscious suppression of individual traits. Repeated interaction with others of similar religious backgrounds has helped define and maintain a distinctive Russian subculture in California's Central Valley (Hardwick 1993). Conforming to a strict set of religious attitudes and beliefs, thus, helps to maintain a distinctive sense of Russianness in this emerging homeland.

Three geographic themes provide evidence of the intense bond with other Russians and with place in the new Russian homeland in the United States:

(1) the establishment of strict property boundaries as perceived barriers to out-siders; (2) exclusivity and territoriality; and (3) symbolic connections to the Sacramento and San Joaquin rivers.

Maintaining Ethnic Boundaries

Throughout the region, the Russian identity is expressed in subtle, unob-trusive ways in the landscape, even though many historic commercial es-tablishments like Petruska's Speakeasy, Nazaroff's store, and the Cossack restaurant are gone. The defining elements of the Russian cultural land-scape in California are residential front yard fences and barrier-style hedges (fig.13.3). An analysis of Russian-owned homes from Chico to Bakersfield revealed that almost every house has solid fencing and tightly closed gates in the front yard (Hardwick 1986). Most house-scapes also have thick cypress tree "hedges" planted behind the fencing, a plant common in virtually all Russian churches and cemeteries. This subtle Russian ethnic signature found at homes, churches, and commercial establishments is evidence of the Russian presence in central California and declares the region's unique ethnicity to those who understand it.

When asked about the fencing, all Russians interviewed expressed surprise to be asked. It seemed perfectly normal for them to build fences and plant cy-press hedgerows, even before their houses had been completed. According to one local resident: "We Russians always have fences or hedges in our front yards. It is our tradition to mark off our own territory. We always lock our front gates, although the Portuguese and Italian people here do not. We just don't understand them at all" (Lokteff 1984).

Exclusivity and Territoriality

The importance of property boundaries and property maintenance expressed in barrier-style fencing so common in Russian neighborhoods is also reflected in home ownership patterns. Owning one's own home and marking off one's territory are important in Russian culture. Ownership of property was practi-cally unheard of among any but the upper class in Russia and the former So-viet Union. As a result, Russians today are very protective of their property and their yards and believe strongly in home ownership if at all possible. A new

Fig. 13.3. Cypress trees and fences form an impermeable barrier between the home and street of Russian-owned properties in West Sacramento, California. Photograph by Susan W. Hardwick, 1997.

Russian resident in central California remarked about home ownership: "My parents bought our land on Riverbend Avenue in 1913. They saved their money, then bought the house. They always dreamed of raising vegetables in their own yard" (Planteen 1985).

More than three hundred Russians interviewed had strong opinions about home ownership. They often expressed personal disdain for "those renters" in their neighborhood. "Good people own homes, bad people rent" is the dominant attitude among Russian residents in central California communities. This common perception provides yet another example of their shared attitudes and values and further solidifies their inter- and intraethnic community bonds.

Holy Waters: Russian Symbolic Bonds with the River

Strong emotional connections with the waters of the Sacramento and San Joaquin rivers provide an ongoing and significant bond and source of attachment for Russian residents of the region. Forming the northern boundary of their ethnic node in West Sacramento, the Sacramento River has symbolized the new homeland since the earliest days of Russian settlement in California (fig. 13.1). Almost every person interviewed mentioned his or her proximity to the river as "the one best thing about living in the Central Valley." Residing near a waterway afforded early residents the traditional pleasures of hunting and fishing and served as a powerful symbolic link for people and place in Russia and the former Soviet Union.

"This river looks *exactly* like the Volga River in Russia" was repeated over and over in many different ways. Although the majority of Russians now living in California originated in Siberia or central China and never saw the Volga, virtually all immigrants know songs and stories about this major river. Well-known and often repeated Volga lyrics resonate with themes of life and death, resurrection and rebirth:

> From places far away, the river Volga flows
> The river Volga flows, it flows on endlessly
> Through fields of golden grain
> And 'cross the snowy plain
> The Volga flows and I am seventeen.

> And mother said, life can bring many things
> You may grow weary of your wanderings
> And, when you come back home at journey's end
> Into the Volga waters dip your hands.

This well-known Russian folk song has strong emotional significance for the Russian people in central California. Baptism in the Volga, representing the connection with "Mother Earth," is a vital part of a deep Russian attachment to their place of birth so very far away. In addition, rivers in the former Soviet Union have long been their only "pathway to the sea" and thus represent an important economic as well as a physical feature of the landscape. The Sacramento and San Joaquin rivers, then, serve as important symbols of the historical linkages with Russia for immigrants now living in California. The connection makes it possible for them to feel "at home" even though they are thousands of miles from their original cultural hearth. Comments from a few interviews express this sentiment and symbolism best:

The river was always the best thing about our house in Bryte. It flowed right by the home we built on Riverbend Avenue (Planteen 1985).

That was before the levee was built and what a perfect place for a Russian to be! The flowing Volga couldn't even have been a better place to be (Vesselerova 1985).

We had our house near the big river in Harbin Manchuria, too. Rivers are very important in our heritage, that's for sure (Karakozoff 1985).

There didn't use to be a levee on the river and it was like being at a summer camp. We could swim everyday and this was a big deal. It might sound strange but it was just so beautiful (Domasky 1983).

Conclusion: Future Patterns, Enduring Connections

Like all cultural groups, Russians in California tend to view their home-space as the center of the universe. Tuan (1977, 154) found the profound attachment to homeland to be worldwide: "It is not limited to any particular culture or economy. It is known to literate and non-literate peoples, hunter-gatherers, and sedentary farmers, as well as city dwellers. The city or land is viewed as mother and it nourishes."

These shared and deep attachments to memories of Mother Russia, the river, ownership of their own land, and to each other bind Russian immigrants and their children to their emerging California homeland. "The affective bond between people and place" (Tuan 1974, 4) in this emerging homeland has been forged through the decades of the twentieth century and no doubt will intensify in the new millennium.

Despite the diversity of socioeconomic and religious backgrounds, scattered distribution of specific places of origin in the former Soviet Union, and varying social and economic backgrounds, a tight-knit sense of "Russianness" exists in the region. What makes this emerging ethnic homeland unique, in fact, is that immigrants and their families have held onto a belief system and set of cultural values long gone in their original home-space. California has become, in a very real sense, an ethnic museum landscape, preserving a culture now almost completely obliterated by political and economic change in the "new" Russian federation. Features of a lost past remain in the context of this new land as Russians in California cling to the old ways and resist change, perhaps knowing they may be the last group to carry them on.

What drew Russians to the Central Valley in the past holds them here in the present: religious freedom, economic promise, and a sense of community. Ongoing emotional linkages to the Sacramento and San Joaquin rivers, feelings of safety and security inside well protected, solidly fenced, owner-occupied properties, and strong connections to religious communities and networks both in California and in their original Russian homeland via radio broadcasts contribute to a vigorous emerging Russian homeland in central California. New and old Russian immigrants feel good about their new homeland, and, according to post-Soviet immigrant Nadia Bolyshkanov (1998): "We hope to keep it as Russian as possible for a long, long time."

CHAPTER FOURTEEN

Montana's Emerging Montane Homeland

. . . .

John B. Wright

Montana is a place that never was but always is—a fractious, undecided landscape where the essential point of living is to discover what kind of homeland it should become. Montanans share a visceral certainty that they are unique but do not agree on what they are; they live in a culture where, to paraphrase historian K. Ross Toole, "optimism outruns the facts" (Toole 1959, 247). There are at least two Montanas, one based on destructive exploitation and transience, the other grounded on stewardship and a fierce devotion to life spent outdoors. The state embodies both intentions of the essential human mystery. This dilemma is played out vivid and real, like two strands of rope tossed into a turbulent Big Sky.

No geographer has ever specified a "Montana" region in a classification scheme. Wilbur Zelinsky referred to "the problem of the West" and relegated Montana to a generic category called "The West," drawing no distinction between Montana and the Mojave Desert, the Nebraska Sand Hills, or the shortgrass prairies of Kansas (Zelinsky 1992, 129). Donald W. Meinig (1972) fared no better in his classic analysis of "American Wests." Other than a concession to Butte and Helena as minor cities, Montana was again cast into the bin of geographic miscellany. In the absence of a primate city, the meaning of the place has proven to be elusive. Joel Garreau lumped Montana with other little-known and thinly peopled lands in his "Empty Quarter" region (Garreau 1981). For geographers, Montana is the place no one knows.

Historians provide a bit of solace. Frederick Jackson Turner (1920) defined a frontier based on low population density and free land, Walter Prescott Webb (1957) perceived a climatological West ruled by aridity, and Patricia Nelson Limerick (1988) revealed a West linked by an unbroken legacy of conquest. However, none discussed Montana in depth nor offered finer distinctions based on emerging cultural geographies.

Montana's historians have. Joseph Kinsey Howard's *Montana: High, Wide, and Handsome* (1943) remains the finest book of its kind. Howard grasped the singularly confused nature of Montana, writing that its history "has been bewilderingly condensed, a kaleidoscopic newsreel, unplotted and unplanned; that of other states has been directed, molded by tradition into a coherent and consistent drama" (Howard 1943, 3). Howard wrote that "Montana never has had a stable economy. . . . [It] is a country of great intensities [where people] are a cash crop" during frequent economic recession. "What are we then?" asks K. Ross Toole in *Twentieth Century Montana: A State of Extremes* (1972, 287). He answers plainly: "This 'thing,' this 'place' called Montana has been cyclically beaten, battered, and bruised. It has often been misgoverned, exploited, lied to, and lied about." Destructive exploitation is the defining characteristic of the Montana experience. The place has been treated as a resource colony and as a "plundered province" (DeVoto 1934, 355). This history has left psychological and geographic scars that go far in explaining why a healing homeland impress has yet to be created.

Few places in America have been treated with more vicious disregard than Montana. Operating more on myth than geographic truth, the exploiters have inflicted or learned desperate lessons. First came the open-range ranchers of the 1880s, overstocking semiarid prairies with 600,000 cows in a country, at the time, incapable of supporting 100,000. When the blizzards came in the winter of 1886/87, the grass was gone and livestock losses in some areas exceeded 95 percent. Following this disaster, barbed wire fenced the land into permanent mixed-stock farms; Terry Jordan calls the pattern the "Midwestern ranching system" (Jordan 1993, 267). Chastisement replaced conceit in the ranchlands of the state.

Then came the "Honyockers" (German slang for "Chicken Chasers")— greenhorn farmers lured into the state by the passage of the Enlarged Homestead Act of 1909. This act, rather than any bill passed in the nineteenth century, initiated Montana's largest homesteading influx. Some 85,000 people arrived in the state's northern plains alone, enticed onto 320-acre tracts by rail-road company promises of agrarian plenty. From 1910 to 1916 the rains came, and wheat prices stayed high because of wartime demand. Then World War I ended, prices collapsed, and drought returned. By 1920, 65,000 settlers in Montana went broke and fled the state (Toole 1959, 235). They'd been rail-roaded. The farms were reassembled into fewer more durable hands.

Mining, however, has had the most obvious socioeconomic and environmental impact. The massive Anaconda Company tightened a copper collar around the neck of the state for a century. "The Company" mined and smelted metal, leveled forests, owned the newspapers, bribed the legislature, set the wages, murdered union organizers, exported the earnings, and finally shut down, leaving Butte and Anaconda the poorest cities in the state and the largest EPA Superfund site in the country (Wright 1998).

Today, subdivision and residential development are the primary forces in landscape transformation. Media myths of Montana as an inexpensive Arcadian utopia free from pollution, crime, and other vagaries of modern life have brought waves of disaffected Americans seeking a fresh start in what they sense is an authentic place. Two-thirds of Montana's population growth in the 1990s came from such immigration. People are searching for something they desperately need but cannot name. They want to inhabit the *idea* of Montana— a Hollywood "homeland" that rivers run through, where half-mad horses can be calmed by the whisperings of gentle ranching folk. When these delusions are unmasked, few newcomers stay to learn the actual complexities of Montana land and life. The ecological and socioeconomic impacts of this "churning" may be more lasting than any previous process. Montana is a landscape of "sequent exploitation," not sequent occupance; of claiming, not settling; of boom and gloom.

The Emerging Homeland

The struggle over the future of western Montana's cultural landscape defines it as an emerging homeland. It is a voluntary, highly self-conscious, montane region with an identity crisis (map 14.1). There is no dominant ethnic group. German, Scandinavian, Irish, English, Native American, and other nationalities are all well represented in longer-term residents. Chinese, Black, Hispanic, and Hmong people are now arriving. There is no singular cuisine, mu-

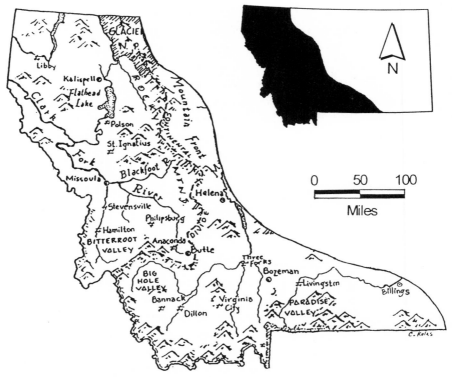

Map 14.1. Montana's emerging montane homeland.

sic, architecture, religion, land-platting pattern, urban form, rural folkway, dress, or political affiliation. The Montana voice is flat with no defining accent, a nondescript "American" sound. The impress of ranching is much the same as in Wyoming, Idaho, or Colorado. Cattle are wintered low and fed hay grown in irrigated bottom lands; in the summer herds are moved onto high elevation rangelands leased from federal and state agencies (Starrs 1998, 158). Other than the "Beaverslide hay stacker" of the Big Hole Valley, no element of material ranch culture is in any way distinctive (Jordan 1993, 305). Grains are grown on irrigated farms using either center-pivot or rolling line systems just as in Texas or Oklahoma (Opie 1993, 122). Dryland grains rise in long strips alternating with fallow lands in a method largely unchanged for a century and widely used throughout the West. Land subdivision plats trace the same lines as those found anywhere. Houses are generic, trailers common, with log homes on the rise, built from uniform sets of plans. Satellite dishes and the Internet erode the former rural isolation. Until 1998, there was no highway speed limit to slow relocation diffusion. Neo-immigrants are rapidly bringing regional, national, and global material cultural to the Big Sky.

Montana's seeming lack of a distinctive ethnic cultural complex (such as in the "Cajun" landscape), or visual signature (such as in the rural "Mormon landscape"; Francaviglia 1970) have led geographers to conclude falsely that there is no true "Montana" region. Yet, essential cultural and spatial characteristics emerge from this seeming geographic monotone only after decades of life spent living in the place. I now explore the often rifted cultural and landscape traits of Montana's emerging montane homeland.

Primary Cultural Traits

Strongly Conflicted "Montanan" Persona

The people of Montana are robustly proud of being "Montanans." There is a shared love of the state that is ferocious and compelling. Yet, there is no consensus on whether the true Montanan is liberal or conservative, pro-extractive industries or pro-environment. The only certain thing is that those who arrive a few years or even weeks after "we" arrived are somehow not "real" Montanans and should stay silent about local matters. There is a widespread cultural amnesia about everyone being immigrants. This perception is particularly ironic, given that in 1880 less than 40,000 people lived in the whole territory.

The cultural personality of Montana is bipolar—a place of vocal, antigovernment militia radicals and silent patriots. The home of the "Freemen" separatists and the "Unabomber" lost more men per capita in World War I than any other state (Howard 1943, 5). The same state that is burdened with a national reputation for harboring neo-Nazis, Freemen separatists, and misogynists of various intensities was the first to send a woman, Jeannette Rankin, to Congress and more recently the first to rise to the defense of Jews. A skinhead threat to a Jewish family in Billings resulted in 30,000 paper menorahs being placed in windows of houses all over the city as a show of support. Neal Peirce (1972, 114) concludes that Montana's politics are "schizophrenic."

Transience

Immigration and emigration are powerful forces in Montana's history. The state's 13 percent growth rate in the 1990s (an increase of 100,000 people) is on a par with percentage rises experienced during the 1950s and 1970s (Von Reichert and Sylvester 1998, 15). Immigration accounts for 66,000 of these new residents. Yet, a net of 51,000 people emigrated during the 1980s (Peirce 1972, 12). This vacillating expansion and contraction is the Montana profile. Transience brings community instability, inefficient and damaging land development, unstable real estate prices, and increasing cultural confusion about the true essence of the place. Through it all, Montanans see themselves as deeply rooted even when this is often not so.

And who are the newcomers? About 40 percent of the amenity migrants come from California and Washington (Starrs and Wright 1995, 421; Von Reichert 1998a, 13). Their mean age is 41 and two-thirds have gone to college, compared with 50 percent of the natives. The average length of stay of all migrants during the decade is just two years. As one amenity migrant leaves another takes his or her place. But some 60 percent are *return migrants*—those either born in Montana who left seeking work or those with family ties in the state (Von Reichert 1998b, 1). Throughout the twentieth century, California and Washington have served as employment safety valves for Montana during economic downturns. This relationship is quickly forgotten when migration flows back to the state. There is a prevailing conviction that "Californicators" are to blame for the development boom despite the inconvenient reality that Montanans are the ones selling the land, houses, and businesses at top dollar to the dreaded newcomers. In Montana popular culture, the Devil displays California license plates. This tightly held conviction exists despite a history in which newcomers are the norm.

Topographic Identities

The state's western half is an archipelago of mountain ranges and river valleys that serve as separate worlds in the day-to-day mental maps of residents. People self-identify as being from "the Bitterroot," "the Flathead," "the Paradise" when referring to the valleys they live in. This physiographic relationship to place is more powerful than county identities despite the use of a numbering scheme on Montana license plates that reveals the home county of the vehicle owner.

Prevalence of Myths

Montanans are, like most westerners, a myth-riddled people (Wright 1998, 19). Some myths are stories that help explain a sacred shared history and the origin of a noble way of life. Others are old fictions reinforcing a tightly held but false geography.

Pristine Landscape. Only new arrivals believe in the pristine landscape until they have lived in the place for sufficient time to see the abundant evidence of mining, logging, and land development. The pristine myth is often brandished by development interests seeking to attract people to housing projects, miners rationalizing the insignificance of their operations, and timber companies overstating the resiliency of forest lands.

Climate. Falsely described as harsher than it is by long-term residents proud of the rigor of the place. Falsely assumed to be milder by newcomers hoping such stories are folklore.

Natural Hazards. The dangers of earthquakes, floods, avalanches, and forest fires are inflated by long-term residents as evidence of their mettle and minimized by new arrivals fleeing the West Coast's array of turmoil. The reality is that real, though moderate, risk exists in Montana for these hazards and more.

Population. Montana's population density is assumed to be low by immigrants. When federal and state lands are subtracted from areal totals, the counties where migrants are moving have densities resembling Hawaii, not North Dakota. Moreover, Montana's population growth rate is believed to be very slow by newcomers and very rapid by locals. The reality is that Montana has averaged 1 percent growth per year since 1950, a doubling time of 70 years (Wright 1998, 38). Although Lewis and Clark explored the region 196 years ago, it is still possible to see many of the landscapes they chronicled. But, if Montana continues to grow at 1 percent per year for the next 196 years, it will have seven million residents, more than the present population of Oregon and Washington *combined*. Such an outcome is not wanted but little dealt with in Montana.

Native Montanans. Being born in the state is a source of robust pride. With it comes an assumption of moral superiority despite the evidence of a full range of political and social views. The indelible myth is that Montana was always a place where "natives" (white people born in the state) were an overwhelming majority and now all the newcomers are destroying that cultural cohesiveness. In truth, in 1920, 30 percent of all Montanans were *foreign born* (Wright 1998, 41). As recently as the 1950s, only 46 percent of Montana residents were born in the state. Today the figure hovers at around 50 percent. Curiously, Indian people are typically excluded from being considered "Native Montanans."

Mining and Timber Economy. Together these two sectors account for less than 3 percent of all jobs and less than 5 percent of the Gross State Product (Wright 1998, 44). Extraction advocates overestimate the importance of these industries to argue for a relaxation of environmental regulations. Environmentalists underestimate the same to argue for cessation.

Cost of Living. This index is 102 percent of the national average in Missoula, Bozeman, Kalispell and other cities where migrants alight. Wages are 21 percent below the national average, ranking Montana 48 out of 50 states in the country. This disparity brings a rude awakening in newcomers and serves as a persistent source of anger in long-term residents.

Crime. Although Montana's rate of violent and property crime is less than half the national average, the state is not the crime-free safe haven sought by migrants and bragged about by residents.

Strong Literary Tradition

The anthology entitled *The Last Best Place* reveals that Montana has one of the country's most diverse and energetic literary traditions (Kittredge and Smith 1988). It is the provenance of writers such as A. B. Guthrie, Ivan Doig, Norman Maclean, James Crumley, Rick Bass, Annick Smith, William Kittredge, Pulitzer Prize–winning Richard Ford, Dorothy Johnson, James Welch, David Quammen, Rick DeMorinis, and Tom McGuane. John Updike has called Missoula "the Paris of the 1990s." The Montana landscape itself is often the leading character in these works. There is a deep and profound appreciation of literature in rural ranching households that rivals or exceeds that of bibliophile urban migrants.

Resource Conflicts

No issue raises more ire in the state than how the land will be used. Given the history of feral extraction in the state, such anger is understandable. It is the home to both herculean mismanagement of resources and enlightened stewardship. In Montana, to paraphrase Shakespeare—the land's the thing.

Montana has the highest coal severance tax in the country: 25 percent of all coal receipts are placed in a fund to mitigate the impact of mining. Mining interests are constantly seeking its diminishment. In 1998, voters banned cyanide "heap leach" mining from the state, mostly because of a threat to the Blackfoot River—Norman Maclean's fabled waterway of "A River Runs Through It." Mining interests have sued to have this law voided. Montana is the only state in the union to have a constitutional guarantee of a "clean and healthful environment" for all citizens. This assurance was expressed in the 1972 Constitution, rewritten mostly to confront strip-mining and increasing development. In 1999, the Montana Supreme Court supported this vague promise with an opinion rendered against mining interests seeking exemptions from review for environmental impact (Montana Supreme Court 1999). True to form, mining companies have mounted a multimillion dollar campaign to change the state's constitution.

Primary Landscape Traits

Power of the Natural World

Despite grim evidence of destructive exploitation, the pulse of elemental Nature remains almost overwhelming in Montana. The state has the most acreage in federal wilderness outside of Alaska, also two national parks, Glacier and Yellowstone. Wild and scenic rivers, national recreation areas, national monuments, and other protected lands are widespread. People report intense physiological responses not merely to the aching beauty but the spirit of this terrain. Unlike most of the country, Montana retains all the creatures present at the time of first European contact. Grizzly bears, wolves, bison, woodland caribou, bald eagles, mountain goats, and the rest of the fauna are being conserved in the face of development forces that would prefer them gone. Directly beside the ruins of failed mining and forestry are portions of wild country that exist in a direct line from the Big Bang. Each day ranches full of wildlife and real proof of stewardship are seen on trips to the grocery store. Nature is still vibrant and ever-present in Montana. It is the most dramatic and obvious trait of this expansive landscape.

Mountains and Rivers

True to the translation of its Spanish name, the emerging homeland in western and central Montana is mountainous (map 14.1). Over 90 percent of the state's population increase in the 1990s has occurred in eight of the state's 56 counties where immigration is strongest: Flathead (Kalispell), Gallatin (Bozeman), Missoula (Missoula), Yellowstone (Billings), Lewis and Clark and Jefferson (Helena), Ravalli (Hamilton), and Lake (Polson) (Von Reichert 1998a, 11). All these settings are beside major rivers, adjacent to high mountain ranges with federal wilderness areas and national parks along the spine of the cordillera.

Population Spatially Compressed in Valleys

As a result of its physiography, and the fact that mountainous lands are mostly in federal ownership, Montana's people are concentrated in linear or compact valleys. In the counties of the emerging homeland, federal and state lands comprise between 60 and 88 percent of all acreage. These government holdings create a spatial compression where residential, commercial, utility, transportation, and industrial land uses vie for open space with agriculture and natural habitat.

Disintegrated Rural Land Development

Significant amounts of Montana's rural agricultural landscape have been subdivided into small tracts. Some of this subdivision dates back to homesteading, some to early land promotion schemes for apple and cherry orcharding, but most of the splitting has taken place in the past 30 years. Rural zoning is essentially absent and other land-use controls are extremely weak. Over 95 per-

Fig. 14.1. Bitterroot Valley subdivision, Montana. Montana's montane homeland is a place defined by intense conflicts between conservationists and developers. This sign in the Bitterroot Valley illustrates the inherent problem—subdivisions are named for what they destroy. Photograph by John B. Wright, 1999.

cent of all subdivision occurs without comprehensive planning oversight or public review, creating a disintegrated pattern of rural and suburban settlement. The texture of the emerging homeland is patchy, marked by chaotic residential sprawl onto agricultural landscapes and a resulting dramatic increase in automobile traffic. Traffic in Bozeman, Kalispell, Missoula, and other "homeland" cities has increased more than 25 percent in the past eight years. Ranching and farming are now sharply declining in economic importance. Yet, over 75 percent of all the vital winter range habitat for big game species such as elk and whitetail deer is found on privately owned valley lands in agricultural production. The effects of land subdivision may prove to be even more indelible than mining. As development occurs, what is threatened is "place" itself (fig. 14.1). Even if people later emigrate, the landscape has been permanently committed to residential uses.

Increasing Presence of Conserved Land

In opposition to development pressures, the six local and statewide land trusts in Montana have conserved over 550,000 acres—the greatest tally for any state. Some 60,000 acres are now being added per year. Open lands under conservation easements and other permanent forms of protection are increasingly obvious. As adjacent and surrounding lands are developed, what remains open is visually striking. In the Blackfoot Valley (near Missoula) more than 25,000 acres within 22 miles of river corridor have been protected from development (Wright 1998). Road signs announce the achievement. Open space systems are expanding in all the cities of the emerging homeland with maps displayed in parks and at hiking trailheads. This landscape indicator of "quality of life" is widespread and expanding.

Mining Destruction-Restoration

All over the emerging homeland, EPA Superfund sites and restoration efforts can be seen: Butte-Anaconda, the Clark Fork River watershed, the smelter complexes in Helena and Great Falls, and others. Much of the restoration remains incomplete. The ransacked landscape of slag heaps, dead forests, polluted rivers, and desertified prairies stands in stark visual opposition to the many wilderness areas in high mountain ranges and ranches conserved by land trusts.

Conclusion

Western Montana is an emerging voluntary homeland defined by conflict over the fate of the land. Wilbur Zelinksy's "Doctrine of First Effective Settlement" does not yet apply in Montana because many claims have been made but none have yet been firmly held (1992, 13). The matter of Montana is far from settled but trends are emerging. The impress of homestead farming is mostly gone; ranching remains but is drawn in two directions, conservation and demise; and mining-devastated areas give evidence of economic failure but the moribund industry that created them is stubbornly trying to reassert its former dominance. Nevertheless, a new, postindustrial culture is now ascendent and environmental organizations devoted to saving the landscape are well organized and increasingly effective. Both Montanas are right before us as obvious divergent choices. Resolving that dichotomy will define the lasting impress of this fractious homeland.

Why an emerging montane homeland in Montana? Because many migrants and conservation-minded locals understand there is still a chance to stem the global march of destructive exploitation. Because you can stand on a rimrock and see a hundred years in either direction. In one way lies a homogenous, depauperate, failed landscape. In the other lies a new land where sustainable occupance serves as a model for other developing regions of the world. The decision is transparently clear in Montana, the native home of hope. Perhaps hope is why diverse people are moving to this emerging homeland and choosing conservation over ruination. In that is reason enough to believe that Montana's "Doctrine of First Effective Settlement" will resolve as one centered on stewardship. Once again Joseph Kinsey Howard (1943, 329) said it best: "The sunset holds infinite promise. Fire sweeps up from behind the Rockies to consume the universe. . . . The Montanan is both humbled and exalted by this blazing story filling the world . . . he cannot but marvel that such a puny creature as he should be so privileged to stand here unharmed and watch."

CHAPTER FIFTEEN

American Homelands
A Dissenting View

· · · · · · · · · · · · · · · · · · ·

Michael P. Conzen

Homelands have made a dramatic comeback since 1990, both in the rhetoric of nationalism and as a concept in academic discourse. The reasons are complex, but the multiple effects of four recent changes in contemporary society appear to have encouraged it. The collapse of Soviet hegemony in eastern Europe and central Asia has opened the door to ethnic self-determination for many peoples once subject to Sovietization. They are now reasserting claims to nationhood and to territory, the latter on the basis of attachment to ethnic homelands of varying ancestry and geographical character (e.g., Kaiser 1994). Then, the defeat of apartheid in South Africa has removed the ignominy, if not the problems, of government-enforced "homelands" designated for blacks in that country (Lipton and Simkins 1993). Further, the fifty-year struggle over Jewish and Palestinian homeland rights in the Middle East has once again reached tumultuous proportions in the tortured search for resolution (Herzl et al. 1989, 16–49; Adler 1997). And, last, the rapid inroads made by global corporatization of economic activity, and the uniformity it brings, together with the attack that modernization has made on traditional lifeways, and the centralization that has accompanied both these processes, have produced counterreactions throughout the world (Poche 1992; Nobutaka 1997). Preserving cultural heritage, especially of groups and regions lacking full autonomy, has become a cause célèbre in many places, and in the articulation of strategies for contesting hegemonic control, the identification of ethnic "homelands" is playing an interesting role.

Whatever geopolitical and commercial shifts the United States has had to make in its international position as the surviving superpower, there have been concomitant changes in coping with its own internal social and political fabric, with respect to ethnoracial makeup, immigration and citizenship policy, and a range of social and economic programs. The transfer of many government services to private supply, reduction in government oversight, and devolution of responsibilities from the federal to state or local levels have combined with major changes in business organization, labor practices, and demographic patterns to produce a new era of cultural competition for limited government resources, access to wealth and power, recognition of "minority" rights, and self-help initiatives in general. The assimilationist model of American society has been seriously challenged for some time, with widespread calls to recognize, celebrate, and sometimes institutionalize on a new scale the nation's tremendous ethnic diversity. American multiculturalism has entered a period in which the Anglo conformity of the past is not just passively resisted but openly contested (Schlesinger 1992). The traditional consensus about what it means, and what it takes, to be American has frayed at the edges, if not at the core.

In such contexts the renewed scholarly interest in homelands, both internationally and domestically, takes on a more than incidental importance. In Europe, the primordial but always socially constructed link between peoples and their ancestral homelands continues to underlie movements for independence or increased political autonomy within the many multinational states of the Continent. In the United States, with a much different ethnographic history, cultural battles and distributional politics appear also to employ strategies and tactics grounded from time to time in group geography. From the land claims of Indian peoples, to the political representational claims of African Americans and the language claims of Hispanics, a diverse set of public issues is bound up with the particular patterns of geographical concentration and dispersion of these and many other groups, mobilized or latent, in the nation's political life (Sutton 1985; Siegel 1996; Shell 1993).

There was a time in America when the idea of *homelands*, insofar as the term was applied to Americans at all, was generally understood in two senses.

In the first instance, there were homelands that belonged to American Indians. Everyone understood that the whole territory of the United States had been the Indians' estate and that over time it was extinguished or reengineered on a smaller scale as formal "reservations," often in new locations (Bjorklund 1992; Frantz 1999). Besides this, homelands were anywhere else in the world, chiefly in Europe or Africa, whence migrants free and unfree had come (Ames 1939; Drachler 1975). Africans captured and removed from West Africa were soon mixed together in their New World locales, producing a syncretic population in the United States in which most individual and group links to particular homelands in Africa were broken. European immigrants in America, on the other hand, often settled in concentrations significant enough to preserve their Old World ethnic ties, especially when refreshed by additional migrant waves well into the twentieth century. Many of these settlements maintained a strong interest in the social and political fortunes of the lands they had left (Vassady 1982; Kivisto 1987).

In the second instance, if the term *homeland* was applied domestically to any area or people aside from American Indians, it was—casually and incidentally—to locales in which people of no special ethnicity had made their homes and from which they derived an obvious regional affiliation, for example, Southern Appalachia, or the Middle West (Campbell 1921; Thurston and Hankins 1954). Such usage was fashionable in educational materials shaped by the old maxim, reaching back to Pestalozzi, that children's learning starts close to home and broadens out from there. Religious missionaries in the nineteenth and early twentieth centuries divided their efforts between missions in foreign lands and those in the homeland, viewing the latter in a clearly national sense (Limouze 1939).

The retrieval and avid study in the last third of the twentieth century in the United States of a wider set of cultural histories than before of particular ethnic and regional groups, however, has set the stage to employ the homeland concept with unprecedented liberality. It is being proposed that diverse groups possess ancestral homelands in America (Nostrand and Estaville 1993). The collection of studies to which this chapter is appended represents a further cross section of such work. Despite efforts to harmonize understandings of the homeland concept, ambiguities abound, permitting like and unlike to be presented under the same rubric. The aim of this essay, then, is to reflect on the basis for recognizing homelands in the United States, especially in light of its meanings elsewhere, and to offer some observations on the clarity or otherwise of several cases that have been, or might be, suggested as American homelands.

The Concept of Homeland

As with any broad social construct, the idea of homeland has both individual and collective meaning. Every individual has a homeland, in the sense of the "land of one's birth," or of one's remembered upbringing. Whether that connection is important depends on the individual's rational and emotional makeup as well as his or her life experiences and social allegiances. But in general its value lies in providing a sense of security and a sense of belonging. In considering where the feeling is most likely to be found, Yi-Fu Tuan (1977, 156) suggests that "profound sentiment for homeland has not disappeared, it persists in places isolated from the traffic of civilization." As the reflection of a universal human trait, however, he notes, "the rhetoric of sentiment barely alters through the ages and differs little from one culture to another." Collectively, on the other hand, the rhetoric of homeland has varied widely in specific content and expression from place to place and over time, reflecting differences in societal organization, political status, and cultural tolerance.

Any discussion of homelands in America needs to begin by recognizing how atypical the United States is in the world, as an immigrant state, in the fundamental relationship of its people to their form of government and territory. Like the relatively few other immigrant states, such as Canada, Argentina, Brazil, Australia, and New Zealand, localized or subnational homelands are at best cultural anachronisms, not part of the new national myth in the making. By contrast, Walker Connor has counted nearly fifty instances where by official state name "the ethno-political myth tends to identify the state politically as the expression or general will of a *particular* ethno-national group and [is] geographically coterminous either with that people's homeland or at least with a segment thereof," as illustrated by states such as Albania, China, Denmark, Germany, Ireland, Israel, Romania, Somalia, Turkey, and Vietnam (Connor 1986, 20–21, emphasis added). Beyond these, the vast majority of cases—constituting, indeed, the global norm—are multinational, multihomeland states, where "the political borders of states have been superimposed upon the ethnic map with cavalier disregard for ethnic homelands" (Connor 1986, 20;

Mikesell 1983). The existence and role of homelands in all such cases is fundamentally different from that found in the handful of the world's immigrant countries, including the United States.

European and Other Contexts

Homeland, literally, means the land one calls home. For the individual this means the land within which one circulates during the course of a life lived generally within a localized community. Before the modern age and the rise of nation states this meant essentially the village domain, as Hobsbawm (1990, 15–16) has pointed out. Only a few merchants, priests, soldiers, and members of the territorial elite ever crossed such local domains. Even consciousness that one belonged to a certain "people" stemmed from one's familiarity with a localized folk community. Thus, the idea of homelands on a scale that came to be equated with nations—requiring a radical feat of imagination well beyond one's own direct experience—is a relatively modern invention. It has served as a tool of nationalism and has become in every way a powerful political idea. Homelands hardly matter in any external sense, until they become the subject of open conflict between neighboring peoples or when captured within centralizing states.

The idea that homelands as regional constructs have primordial status in the evolution of *ethnics* (peoples) has long been problematical. Consensus on when and where the Goths, the Slavs, Celts, or the Germans, for instance, first coalesced as peoples in territories they could regard as their own has proven elusive (Manczak 1986; Golab 1992; Markale 1993; Ehringhaus 1996). The vast geographical movements and ethnic reformulations of these and other groups in Europe and other world regions over the last two or three millennia have greatly complicated the issue. What has emerged today is an established pattern around the globe of between two to three thousand reasonably well-defined regional ethnic communities that exist as nations or protonations, in Connor's terminology, complete with some form of territorial expression (Connor 1978; Cohen 1997, x).

In this global context, the concept of homeland has found more analytical favor among political scientists, sociologists, and political geographers than among anthropologists or cultural geographers (Neumeyer 1992). For the sociologist Anthony D. Smith (1992, 438), "an *ethnic* is defined . . . not only by historical memories, codes and ancestry myths, but also by its possession, or loss, of an historic territory or 'homeland.' Over the generations, the community has become identified with a particular, historic space, and the territory with a particular cultural community." In modern times, with the rise of nation-states, and the instrumentalist purposes they serve, ethnic homelands have inevitably become inextricably intertwined with issues of nationalism and nation-building. In this connection, Robert J. Kaiser (1994) has provided perhaps the most considered exposition from a geographical point of view of the structural relationships and meanings of ethnocultural homelands in relation to nationalism.

To Kaiser (1994, 6–9), a nation is "a self-defining community of belonging and interest whose members share a sense of common origins and a belief in a common destiny or future together." The objective characteristics of nations (common language, land, religion, customs, dress, diet, etc.) have both symbolic and instrumental value. Once national self-consciousness has been created, the homeland serves as the "geographic cradle of the nation and also the 'natural' place where the nation is to fulfill its destiny" (Kaiser 1994, 10). This locus provides not only a sense of common ancestry as a human group but also a sense of a shared birthplace. What gave both of these anchors of heritage effectiveness at the supralocal scale was a "triple revolution" consisting of capitalist economic development and integration over wide areas, the rise of a bureaucratic state apparatus that permitted power to centralize, and a cultural-educational revolution that standardized life sufficiently to render common social experience recognizable over larger territories than before. These changes permitted the myth-symbol complex, grounded in folklore, to become a mass-based entity, by relocating the peasants from a sociocultural periphery to a sociocultural—if not a socioeconomic—core, securing their loyalty by promising a new "golden age" just over the horizon (Kaiser 1994, 12–13).

"The nation's unique history," James Anderson (1988, 24) has written, "is embodied in the nation's unique piece of territory—its 'homeland,' the primeval land of its ancestors, older than any state, the same land which saw its greatest moments, perhaps its mythic origins." This belief reinforces the conviction that the nation is a primordial organism, rooted in a specific place, and for indigenes this creates a strong emotional attachment, a bond stemming from time immemorial, and a belief that they belong there and nowhere else.

Since most populations have not remained stationary or isolated, this raises issues, often furiously contested, about when such sacred attachments first
emerged, involving rival claims over the earliest, longest, latest, or most ben-
eficial period of "ownership." Serious conflict has long been a reality, down to
the present (Kaiser 1994, 18; Knight 1982).

If heritage binds indigenous people to their homeland, so too does a sense
of common destiny. One key destiny is to achieve a modern golden age, for
which demographic regeneration is essential, and the breeding of national
consciousness in each new generation. Hence, history must be reinterpreted
to fit the needs of each new age, incorporating new events and selecting alter-
nate symbols from the nation's past, because the "trajectory" of the future is
anchored in the past, which mediates the dialogue about the future in the present. If common origins promote intranational solidarity, common destiny pro-
motes imperatives for international separation, making the multinational,
multihomeland state anathema. The nation's sense of spatial identity justifies
with history a nationalistic sense of exclusiveness. The homeland is not simply where indigenes feel most at home, it is the place they alone should control, to be "masters of their own land" (Connor 1986; Kaiser 1994, 21–22).

Kaiser regards national territoriality always as a latent sense of homeland,
activated when the nation-homeland bond is seriously threatened, by members seeking to control their own destiny, which they believe would be impossible without a homeland. In multihomeland states, he suggests four main
catalysts: the geographical mobility of "alien" peoples across homeland borders; the social mobilization of indigenes, particularly in the face of increasing
competition with nonindigenes for homeland resources; state-sponsored integration (e.g., Sovietization, Serbianization), coupled with the threat of dena-
tionalization; and, last, centralization of state decision-making (Kaiser 1994,
24–27). Once such catalysts operate, the territorial response in the name of
survival comes in the form of efforts at indigenization, raising barriers to outside infiltration, the ultimate form being political separation, by whatever
means. Separatist movements often secure increased autonomy (e.g., Quebec).
In the wider picture, as Robert Sack (1986, 29) has argued, territoriality tends
to engender more territoriality. In the twentieth century, "it is the nation that
has become the predominant community of interest . . . the collectivity able
to outcompete other groups that may lay primary claim to a member's time,
resources, and loyalty" (Kaiser 1994, 30). In a strikingly apt observation,

Kaiser (1994, 31) notes, "national self-consciousness can and does exist along-side affiliation with several other communities of interest, and it may be difficult to determine whether an individual has attained a national self-consciousness until he or she is put to the test (i.e., 'when the chips are down')."

These theoretical formulations apply with particular force to political and cultural conditions in Europe, but also other parts of the world, over the last two or so centuries. They help account for the vehemence and periodic violence that have surrounded the many challenges to homeland that have been mounted at the regional, national, and international scale, including outbreaks of genocide. In such cases, homelands are ancestral, usually reaching back into the mists of time, socially constructed and consciously manipulated. In fact, they would hardly exist in the group mentality if it were not for their fiercely political and nationalistic purposes. To be sure, homelands in Europe and elsewhere have also disappeared or been merged in the course of long-term ethnogenesis and successful colonizations, but their conceptual salience both for indigenes and for detached scholars resides in the fact that they are tools for political transformation. What, then, of their significance and value in essentially immigrant nations? The United States presents an inviting case for comparison.

American Contexts

While no immigrant nation has developed on land completely devoid of aboriginal habitation, the implication of this type of nation is that conditions of space, resources, and technological superiority have allowed alien colonizers to establish permanent populations in new environments. Are homelands that might be identified in America to be analyzed conceptually on a comparable basis to those in Europe and elsewhere? On the face of it, the answer should be affirmative. If so, then, in such a settler setting, what relation do immigrant populations and those they displaced have to "homelands" defined in the terms discussed above? The first point to be made is that the United States does indeed contain a number of homelands, by any definition.

No one can deny that American Indians, for example, can trace their ancestry through a chain of generations on the continent reaching back well before the first Europeans arrived. Their ancestors were, in relation to any late-comers, the first peoples. While the task of scientifically linking specific Indian tribes of the so-called historic period with earlier tribal cultures of the so-called prehistoric period continues to challenge anthropologists, there is no reason to dispute present-day Indian claims of such antecedence. Nor can one deny that a number of Indian groups have occupied particular areas in the present-day United States for centuries, which they regard as their ancestral homelands (Prucha 1990). Many groups insist on identifying themselves as "nations," to stress the conceptual equivalency between their claims to territory within the present United States and those of Euro-Americans. Other Indian nations shifted their essential territorial affiliations greatly both before and during the long period of European penetration, thus complicating the issue of which groups now have the best historic claims to which localities. Massive population losses, blending and reconstitution of tribal communities (not to mention ethnoracial mixing), and forced movements from accustomed territory have all complicated the prospects for recognizing appropriate homelands on historical principles (Sutton 1985; Ross and Moore 1987). And the cumulative actions of the United States government with respect to the establishment and the undermining of formal reservations over the course of the last 150 years have, of course, added further layers of complexity and ambiguity to the problem (Frantz 1999). Nevertheless, whether historically plausible as ancestral homelands or as largely artificial creations, many of today's Indian reservations function for better or worse as homelands for those groups who resist assimilation with the dominant civilization.

It is also fair to say that, since the political founding of the United States in 1789, the American nation has been engaged in developing its own homeland. For Americans this ongoing project supersedes in formal and sociopolitical significance all other territorial facts, except for recognizing the treaty homelands of the Indians. For people who simply consider themselves Americans, there is only one homeland in North America, namely, the entire United States. (For those who lived through the Vietnam war years of the 1960s, the anti-antiwar slogan "America—love it or leave it!" has a familiar nationalistic ring.) Given the general diversity and the recency of some populations comprising the American people, and the comparative recency of the nation's completed geographical extent, this homeland is in many ways qualitatively, if not conceptually, different from homelands in regions where ethnic groups have been residentially distinct for many centuries, as in Europe. Time has simply had less opportunity to etch the markings of homeland into the landscape and the psyche of the inhabitants. Within the American national homeland, of course, patterns of local affiliation, affection, and loyalty can be found in marked degree, but these are for the most part attachments to regions (regions filled nonexclusively with numerous ethnic and cultural communities) contained within a larger unitary entity, not bonds to a primordial, ancestral homeland of a separate "people" (Zelinsky 1988).

The essential argument here is that homelands are the products of indigenous peoples and the crucial repositories of their identity, and as such claim their exclusive or essential loyalty. As an immigrant nation, the United States contains citizens who trace their ancestry back to the Mayflower as well as citizens whose residency might be as short as five years. Thus, Americans as a society have been in the process of becoming indigenous by stages and over most of the national territory, but this indigenization has been uniform over neither time nor space. African Americans in their cultural evolution and settlement patterns, both slave and free, best illustrate this point (Davis and Donaldson 1975). None of the United States' colonization of its national space has been achieved through exclusive ethnic occupation of large areas, prolonged isolation, or endogenic development. There has been too much rapid migration and remigration, too much interethnic and interracial mixing, and too many cultural stimuli from places outside the nation to claim that any ethnic subgroup within its confines and since its inception has had the opportunity to create an ancestral homeland, except—slowly and in fits and starts—at the national level.

The comparative recency of American territorial completion, in fact, suggests a means of distinguishing conceptually between plausible and implausible cases of homelands within the United States. It might be expressed as a theorem: present or remnant homelands in America can be found only in zones that lay beyond the political domain of the United States at the time of their maximum florescence. As American national control was extended and took effect, any prior ethnocultural homeland would encounter strong pressures for dissolution, and no excuse would exist for the development of new subnational homelands. This idea does not mean that preexisting homelands would immediately collapse, but at best they would languish and mostly fade away. We shall return to this idea later.

Conditions for Homelands in the United States

Since the emergence of geographical interest in homelands in the United States a variety of criteria have been proposed for recognizing them—some simple, some complex; some hierarchical, others not; and by no means all compatible with each other (Carlson 1990; Nostrand 1992; Nostrand and Estaville 1993; Roark 1993; Conzen 1993; Jordan 1993a; Lamme and McDonald 1993; Sheskin 1993; and Boswell 1993). In the enthusiasm to discover additional examples beyond the well-researched Hispano case, what constitutes a homeland has clearly varied with the eye of the beholder. A preliminary issue arises over the question whether a homeland should be seen as a special case of the anthropologist's and geographer's "culture area" or "culture region," or whether it is a separate category altogether. Opinion still varies (on this distinction, see Conzen 1993, 14–18; for the former view, see Nostrand and Estaville 1993, 3; for something approaching the latter, see Jordan 1993b, 75). Still other opinion remains largely unpersuaded by the rush to declare the need for the term (Zelinsky 1992; Zelinsky and Lee 1998).

What Criteria Should We Use?

Richard L. Nostrand and Lawrence E. Estaville Jr. (1993, 1–3; 1995), both in their earlier calls for examination and in the current collection, stress five "ingredients" that must be shown to operate: people, place, time, control, and bonding. With regard to people, they suggest that some ethnic groups, some folk cultures, and some popular cultures might qualify. Significant is that the groups need to be "self-conscious people." Regarding place, the territories should cover substantial parts of at least one state, in which the people have adjusted to their environment with a "unique cultural impress." On the matter of time, they envision no set minimum period, leaving the question completely open. Their notion of control covers "some measure of political control," which might be based on simple demographic strength (winning local elections, and land ownership, for example). The "key" ingredient is bonding. They argue that a developed sense of place produces "feelings of emotional attachment and identity, desires to possess, perhaps even compulsions to defend." It is the uncommon degree of people-place bonding that differentiates homelands from folk culture regions. "The homeland concept, it seems to us, is rooted in human ecology: it is a special type of culture area or region because of people-place bonding" (Nostrand and Estaville 1995).

Difficulties arise over the vagueness of some of these ingredients, and the means of measuring them in a strictly comparative way. In many instances, these group characteristics are most appropriate in defining cultural groups that comprise classic culture areas or culture regions, and it is not clear what marks them out as supposedly more potent homelands. What are the necessary thresholds of self-consciousness that distinguish a homeland-qualifying group from one that merely inhabits a culture area? What "unique cultural impress" is of a higher order of uniqueness than that of a group dwelling in a culture area? Does it have to do with survival of more traditional modes of livelihood, less modern and commodified? Long spans of time for cultural development seem self-evident, but in setting no minimum period can the time required in some instances be so short that first generation settlers automatically represent sufficient ancestry in the area in question? The attention to political control is minimal. It could be suggested, however, that the issue here is not so much winning local elections or owning parcels of land, which all prominent groups do in their respective culture areas—it is outright autonomy. Individual states within the federal union do not have absolute control over their population's lives, because the federal constitution substantially limits a state's power. No American state area coincides with any proposed homeland, making the prospect of meaningful cultural autonomy even more remote. With regard to bonding, the weakness lies in giving salience to ecology rather than to politics. Which regionally settled cultural groups in America, one might ask, are *not* by now reasonably well adjusted to their physical resource environment? Is it not more pertinent to ask, which groups are devoted well beyond others to preserving their "peoplehood" and a way of life in their territory, and are willing to battle for full or near-full political autonomy to do so? These are the standards for gauging the strength of homeland among states and substate areas elsewhere in the world.

Other contributors to the debate have added further considerations. Michael O. Roark (1993, 6) places special stress on common ancestry, exclusivity of possession, and group identity with territory, involving even a sense of nationalism. But he makes a special plea for the "emotional tie or bond between a people and the land," without suggesting in an American context how one proves, as a practical matter, that "my people love their land more than your people love theirs."

Terry G. Jordan's (1993b, 75) definition also emphasizes a self-conscious group "exercising some measure of social, economic, and/or political control" over its region. He goes on to say, "Its closest analogy is the nation-state, but homelanders lack and generally do not seek independence. Homelands are incompletely developed nation-states that give expression to sectionalism." The analogy with nationhood is appropriate, but the inference to be drawn is that peoples with meaningful homelands, in the nationalistic sense, do indeed seek independence, and when you have no urge for independence you have at best a very weakly developed sense of homeland, perhaps comparable to the more general "love of place" that Tuan finds to some degree in most individuals and groups (Tuan 1977). Sectionalism, a term popularized by the American historian Frederick Jackson Turner to describe regional political differences in the United States, translates elsewhere into secession movements (whether successful or not) or simple regionalism (when external hegemony prevails). Sectionalism in the United States only once led to short-lived secession, and not for differences of fundamental peoplehood but rather a way of life based on a particular form of labor exploitation. (That Robert E. Lee was at one juncture asked to command the Union armies reminds us that he was not regarded as an alien by northern interests in this interregional conflict.) Since modern nationalism emerged in the nineteenth century, homelands are indeed a key basis for national striving, as Kaiser has shown. They may be incomplete nation-states, but not for lack of interest in independence if given the choice. Among American regions, entities akin to "incompletely developed nation-states" more likely reflect fatally incomplete homelands, thwarted in their development by the commonalities infused by a spreading United States nationalism.

Axioms for Defining Homelands: Three Dimensions, Nine Criteria

This line of argument suggests that there is an imbalance and an incompleteness in the currently fashionable working definitions of homeland being employed when scouring the United States for evidence of homelands. These have to do with possible overemphasis of ecological adjustment as a discriminating variable, underemphasis of the political-historical context of regional consciousness, and the dangers of special pleading with respect to such senti-ments as "attachment to land" or even attachment to group. The problem is the more difficult because very few participants in the debate who have sought to present detailed evidence for past or present homeland character have so far worked on more than one cultural group. Thus, few have grappled with the direct task of objectively comparing subjective elements of different cases. For those interested in conceptual models and analytical methods that permit comparative work to be done, it is important that work proceed on compara-ble lines. Although the writer has offered an earlier general discussion of homeland definition (Conzen 1993, 21–24), it appears upon reflection, and particularly in light of the comparative literature, that a reworked set of es-sential criteria should be drawn up. Three dimensions seem crucial for a def-inition that can be applied systematically anywhere, and each of these dimen-sions would appear to offer means of measurement in three distinctive ways. This new set is clearly not expressed in fully operational terms, but it is offered as a more conceptually robust scheme than what has gone before. The key components of homeland recognition are:

IDENTITY

1. Ethnogenesis: a sense of peoplehood
2. Indigenization: time to develop in place over multiple generations
3. Exclusivity: promoted through geographical isolation

TERRITORIALITY

4. Control of land and resources
5. Dedicated political institutions
6. Coherently manageable spatial unit

LOYALTY

7. Defense of the homeland against "alien" intrusions
8. Compulsion to live within the homeland
9. Production and veneration of "nationalistic" landmarks

Identity. Most important for identifying homelands in continental America is to establish which groups occupying territory perceive themselves to be a dis-tinct people. This distinction should be globally relevant. In order to consider what distinctions could be analytically useful, we might note a recent scientific classification (O'Leary and Levinson 1991, xxiii). Anthropologists consider the cultures of North America to fall into three broad categories: Native Americans (109 separate cultures within the 48 states), folk cultures (17), and ethnic groups (at least 48 cases). Thus, the coterminous United States alone contains within its borders at least 174 distinct cultures, some extremely small and highly localized and others large and widely distributed. A tally by historians of American ethnic groups counted 106 cases, though considering all American Indians as a single group (Thernstrom 1980, vi). Without doubt major variations exist in the self-perceptions of these groups, and determining such differences systematically at this point is an impossibility. Three criteria for relating group identity to the issue of homeland emerge from discussion so far.

1. There should be convincing evidence that a cultural group is a *distinct people*, based both on objective and subjective measures, and not the product of recent "export" from a homeland elsewhere. Setting aside the problem of evidence for the subjective dimension, we must look for objective evidence that the cultural group has undergone significant ethnogenesis during its time in America, either becoming a hybrid people through the blending of different groups or through substantial endogenic culture change, including emphatic divorce of the diasporic group from the mother culture through rejection, neglect, or divergent development. Either way, this creates a distinctly new symbolic and material culture system, not necessarily losing all the characteristics of previous pools, but with evidence of new elements not present in the previously separate groups or at least not central to their culture systems.

2. Closely related to ethnogenesis is the issue of indigenity. *Time* is the essential factor here. At what point do cultural groups, assuming they stay put, become indigenous? If they have been present "from the beginning," their status is beyond doubt. Regardless of whether, as latecomers, they have colonized open and available land or appropriated land and resources from expelled populations, or if they simply have settled in the midst of existing groups, how and when do newcomers establish their own indigenity? The longer a group has time to develop, the more nearly indigenous it becomes, and the more past generations accumulate a record to give the group a sense of American ancestry. Several generations seem minimally necessary.

3. Equally important to indigenization is *geographical isolation*. If a group develops in geographically isolated circumstances, its endogenic development is encouraged, pressure for assimilation with neighboring or commingled groups is avoided, and exclusivity is promoted. Perhaps it is axiomatic that dis-

tinctive cultural identity has been best preserved in America in isolated locales over long periods of time; absent these conditions, cultural fusion will likely sap group identity. (The Amish have resisted absorption at close quarters, but they are a rare exception.) It might also be suggested that in the wide spaces of America extreme isolation has sometimes made up for the otherwise insufficient time in which "new" groups have emerged.

Territoriality. Besides clear cultural identity, there needs to be evidence that a group has over time appropriated a territory to which its identity is attached, a theater for the collective memory of its cultural achievements, a perpetual source of livelihood and renewal. This territorial dimension can be assessed by three further criteria.

4. To lay claim to a homeland a cultural group needs to exercise exclusive or preemptive *control over land and resources.* Homelands are not by their very nature to be shared by more than one group, so exclusive control over territory is the objective. Such control allows not only freedom to exploit the land but also to define such other issues vital to group identity as migration (who is allowed in) and citizenship (who is allowed access), taxation, language, education, religion, and so forth.

5. This requires the *formation of dedicated political institutions* within the homeland designed to defend the people's heritage and develop their independent destiny. Before nationalism, ethnocultural groups sought expression within the context of kingdoms and empires, and localism was as much a defense for as well as a limitation on culture-building. But with the rise of mass societies national institutions have provided the most potent forms of cultural shaping. Localism and regionalism have supplied considerable opposition to nationalism, but once firm central authority has been established and its institutions diffused throughout a would-be homeland the larger influence is hard to counteract. American nationalism, armed with "Manifest Destiny," proved a difficult juggernaut to derail, and, as the Mormons discovered, competing political institutions stood little chance of finding firm root in what became American territory. Necessary to any plausible homeland, then, is the commitment to create and maintain homeland institutions dedicated to political autonomy, controlled by homeland members in order to fulfill their destiny as a people.

6. A sixth criterion by which the evidence for homeland formation can be assessed is the extent to which a cultural group can mold its territory into a *vi-*

able and coherent spatial unit. This requisite is more than simply abstract control: it is the ability to define and manage the periphery and rationalize the interior in such a way as to maximize the homeland's long-term sustainability and ease of internal communication, because both symbolic and pragmatic mobilization depend upon these features. Size (to engross land and mineral wealth sufficient to the needs of the people in question) and compactness of form (for ease of accessibility and administration) are obvious attributes that support this objective. Significant is the extent to which the territory controlled is coterminous with the distribution of the "people." Similarly, too much fragmentation removes any basis for homeland coherence (as Pakistan found out in losing Bangladesh).

Loyalty. Last of the three fundamental dimensions is the role that subjective attachment to the heritage and destiny of a homeland plays in confirming this type of geographical region. Loyalty is not simply the existence of a generalized group "sense of place" or "love of place." It is rather an imperative to defend the territory because by so doing, and at times only by so doing, is one able to defend the culture itself. This loyalty takes three geographically interesting forms.

7. Evidence that members of a cultural group are *willing to defend the homeland* against outside attack or infiltration makes a good case that loyalty to place and group actually exists. What form that defense takes, and who are regarded as allies and enemies are critical signs of the character of the loyalty at issue. Sustained political action on an unceasing basis may be required. In a dire emergency, it may require a willingness to die for the cause. People the world over pay with their lives to save the homeland. In an American subnational context, how often has this been the case? At the national level, certainly, many in this century died to save New Mexico from fascism and communism, though it took them across the world to do so. But they did so in the name, not of New Mexico, but of the United States, the national homeland. Now, if people are not willing to die for their subnational homeland, what hold does it have over its adherents? Patriotism is different from nostalgia. The issue is on what level, when the chips are down, does the primary loyalty—and territorial consciousness—lie?

8. Along with defense of the homeland, there should be evidence of compulsion among members of the group to *live within the homeland* and for "exiles" to return. Economic hard times and severe work limitations might force

temporary relocation, with attendant repatriation of earnings and visits when able. But the test is whether members of the group believe as an article of faith that, when all is said and done, they belong there and nowhere else. What gives homelands around the world such power over people is the conviction that they need to participate in the life of the homeland to achieve their own full self-realization, within the traditions and comforts of a community of belonging. In the absence of a demonstrable pattern of homeland preference, there can be no strong claim that the homeland exerts any effective pull on the group's members, psychic or practical. Arrangements for one's remains to be buried in one's birthplace after prolonged absence only points up the weakness of such homeland attachment in such cases, particularly if the absence has been voluntary, and suggests nostalgia rather than commitment.

9. Last, homeland attachment should yield evidence of active, even enthusiastic, *veneration for the defining symbols and landmarks* of nationhood and peoplehood distributed across its face. Such landmarks are the landscape expression of cultural ideology and traditions purveyed through school indoctrination, and as such suitable subjects for personal pilgrimage. Active veneration should appear in a measurable willingness to pay extra for the continuous multiplication and upkeep of such symbols, because "peoplehood" is an ongoing cultural project. Similarly, if the territory is indeed a homeland there should also be evidence of battles fought to remove public symbols antithetical to the nation's (that is, the peoples') collective image. In this regard, periodic disputes over Confederate monuments and battle flags in the American South offer good evidence that the South is no group's exclusive homeland (Sack 1997).

Short of these nine, rather stringent, conditions being fulfilled, it can be argued that cultural patterning within American space has produced, not homelands, but "culture areas or culture regions" in various configurations of discrete and overlapping distribution, or cultural space of other types. This observation is no demerit. Occupying all or part of a culture region, as opposed to a "homeland," does not invalidate the capacity of a group to display often robust attachment to place. Such sentiment quite likely results from healthy interregional and intergroup competition within the matrix of American social development as a whole. It should be remembered that, conceptually, there is nothing about the culture region concept that excludes or presumes an absence of attachment to place on the part of its residents. Thus, observers need to remain free of the temptation by their own attachment to place or through overly sympathetic study of one group or another to endow their targets with near-mystical bondedness to localities while discounting possibly similar sentiment in others. Many of the so-called American "homeland" identifications may in reality be no more than good exercises in specific culture area and "ethnic substrate" delineations (for this term, see Conzen 1993, 19–21, and below). To restate the issue, what needs clarifying is the nature and strength of the trait "surtax"—the special conditions that apply—that singles out a few cases of culture areas that merit designation as homelands from the rest that do not.

The principal aim in reviewing homeland concepts from an international and interdisciplinary perspective has been to set questions of group culture firmly within a comprehensive framework for analysis. This framework needs to include the political dimension of cultural identity. It is no more helpful to ignore political ideology in constructing cultural-geographical models of homelands than it is to disregard cultural context in political geography. In this case, questioning the degree to which the American nation contains subnations that might consciously seek (or have sought in the past) to establish and defend homelands provides a reasonable basis for distinguishing between cultural space describable in the accepted terminology of culture regions and that requiring the term "homeland."

Types of Ethnocultural Space in America

This is not the place for an extensive assessment of the merits of each of the suggested homelands in the present collection. The criteria proposed in this chapter, however, developed in a comparative context, do invite critical evaluation of the detailed claims. In the space available here suffice it to consider in cursory fashion how the cases pro forma seem to fit the most discussed types of regional cultural space in America.

A summary of the distinct United States cultures listed in O'Leary and Levinson (1991) is presented together with information pertinent to their geographical expression within the national territory (table 15.1). Rounded population figures are from recent sources. External homelands are given where appropriate. Several groups are sufficiently established that they can be considered indigenous, and therefore no longer cleave to homelands outside the United States (former source areas are given in parentheses; a reconstituted homeland, Israel, in brackets). The broad timespan during which ethnogenesis in the United States has been operative is suggested. These estimates do not include the first appearance of group members in what is today the United States; rather, they concern the period of mature settlement and are derived from the ethnographic descriptions in O'Leary and Levinson. In earlier times, most groups were rural in orientation and thus capable of occupying space at the regional scale; in modern times, most immigrant groups are urban-oriented, and occupy urban neighborhoods, often whole networks (archipelagos) of residential districts both within individual urban areas and among selective networks of cities. The regional cultural space of all these groups, in my judgment, falls plausibly into five general types: homelands, culture areas, substrates (or enclaves), islands (and archipelagos), and urban space (definitions below).

As a class of space, cities are poor locales in which to seek homeland dynamics, except as articulation points for the political mobilization of cultural self-determination in the surrounding region. In this respect they can, with varying degrees of ethnic autonomy, act as cultural capitals (for example, San Antonio, discussed in Arreola 1987). But as often, cities have been, and are, tools of empire and subjugation, key venues for challenging the homeland sentiment among subordinate peoples with the symbols and institutions of the dominating culture. Such an outcome is markedly so in the United States, and therefore, cities have most relevance in considering the advancement of the American national homeland, an argument that presents complexity enough for another study.

Homelands

There is a fundamental difference between saying "my (or his, her) homeland" and "the homeland" or "a homeland." The former identifications refer to place associations of individuals, who all are entitled to consider some place "where they are from." On the other hand, *the* or *a* homeland refers explicitly to a group's homeland, so the identity of the group as a collectively self-conscious "people" with its own distinctive heritage and cultural mission (not its individuals) becomes the key cultural unit of study. In thinking about homelands in America, it seems that the plausible groups and their areas should score well on all the nine criteria identified in the preceding section. Any serious deficiency or absence of evidence in a category might suggest that the case belongs

TABLE 15.1. *Anthropologically Distinctive Cultures in the United States, and Their Characteristics Relevant to Cultural Space*

Group	Culture Categories	Population (incl. ests.)	External Homeland	Ethnogenesis (yrs.)	Predom. Cultural Space in the U.S.
AMERICANS (all citizens)	MC	265,000,000	—	225	Homeland (national)
AMERICAN INDIANS	NA	2,000,000	—	Primordial	Remnant homelands (Reservations)
African Americans	E-2	31,000,000	(Africa)	300	Substrate
American Isolates	E-4	?	—	Var.	Islands
Amish	F-1	130,000	(C. Eur.)	200	Archipelago
Appalachians	F-2	6,000,000	—	150	Culture area
Basques	F-1	40,000	W. Eur.	100	Islands
Black Creoles	E-4	80,000	—	200	Substrate
Black West Indians	E-3	220,000	Carib.	80	Urban
Cajuns	F-2	800,000	—	200	Remnant homeland
East Asians	E-3	3,000,000	Asia	100	Urban
EUROPEAN AMERICANS	E-1	204,000,000	Europe	Var.	Substrates, islands
Haitians	E-3	800,000	Haiti	25	Urban
Hasidim	F-1	100,000	(E&C Eur.)	50	Urban
Hutterites	F-1	31,000	(E&C Eur.)	100	Archipelago
Irish Travelers	F-1	4,000	Ireland	150	Itinerant
Jews	E-3	6,000,000	[Israel]	150	Urban

			Mex., etc.	Var.	Substrates
Latinos	E-2	21,000,000	Mex., etc.	400	Remnant homelands
Hispanos	E-2	700,000	(Spain)	150	Enclave
Tejanos	E-2	870,000	Mexico	200	Islands
Lumbee	E-4	40,000	—	300	Dispersed, urban
Mennonites	F-1	200,000	(Europe)	50	Islands
Micronesians	E-3	60,000	Pacific	90	Urban
Molokans	F-1	20,000	Russia	150	Substrate
Mormons	F-2	4,000,000	—	80	Dispersed
Old Believers	F-1	7,000	Russia	150	Culture area
Ozarks	F-2	2,000,000	—	40	Urban
Polynesians	E-3	60,000	Pacific	80	Itinerant, urban
Rom (Gypsies)	F-1	20,000	S.E. Eur.	300	Archipelago
Sea Islanders	F-1	n.a.	(Africa)	200	Island
Shakers	F-2	12	—	100	Urban
S. & S.E. Asians	E-3	7,000,000	Asia		

SOURCES: O'Leary and Levinson 1991 (*NA, F, E); author's calculations (all categories, plus *MC).

MC = Macroculture; NA = Native Americans

F-1 = Folk culture, developed elsewhere and maintained as a distinctive culture and identity in America

F-2 = Folk culture, developed in America

E-1 = National-origin ethnic group (formed and maintained through group members sharing a common affiliation based on ancestry traced to the same nation)

E-2 = Ethnic minority group (like E-1, based on members affiliating according to national or regional origin; but also object of economic, cultural, or political discrimination based on race or religion)

E-3 = Immigrant group (recently arrived and largely unassimilated)

E-4 = Syncretic ethnic group (blending of people and cultural features of two or more distinct groups)

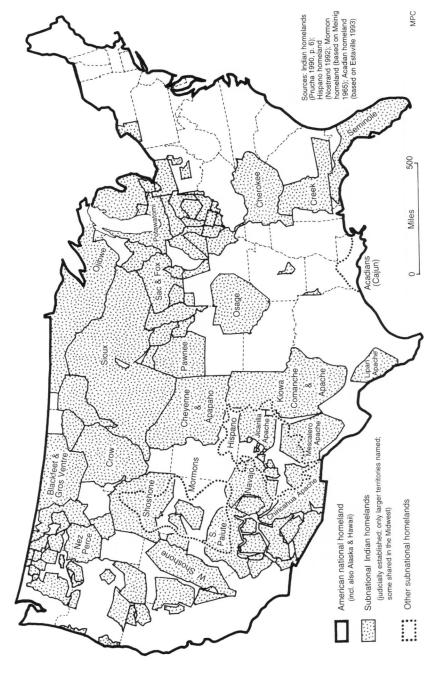

Map 15.1. Selected ethnocultural homelands in the United States.

Sources: Indian homelands (Prucha 1990, p. 6); Hispano homeland (Nostrand 1992); Mormon homeland (based on Meinig 1965); Acadian homeland (based on Estaville 1993)

MPC

American national homeland
(incl. also Alaska & Hawaii)

Subnational Indian homelands
(judicially established; only larger territories named; some shared in the Midwest)

Other subnational homelands

0   500
Miles

to one of the other types of cultural space. Homelands, or traces of former homelands, within the United States imply distinct, not yet fully assimilated "peoples" within the national territory. Assimilation, of course, does not mean loss of all distinctiveness—regional differences of one kind or another almost always survive even, or especially, in the larger and more complexly "assimilated" societies. That some homelands undoubtedly existed was certainly true in the past for a few groups under special geographical circumstances (isolation, distinct ethnogenesis) and still formally applies to many American Indian peoples (map 15.1).

The American National Homeland. Before reviewing subnational cases, it is worth stressing the case of the American national homeland, particularly because it has received no separate chapter treatment in this collection. There is, to be sure, one large, vibrant, continuing homeland in America—the United States itself. It is a conscious creation, now over two centuries old, and as a formal political and cultural entity the most open and inclusive of any known homelands. It serves successfully as a territorial expression of a national community possessing a strong sense of peoplehood, regardless of the multiple identities and social divisions, ascendant and declining, that coexist within it. In this respect American identity is generically no different from ethnic identity, though it is more inclusive, and pursues the most ambitious of cultural projects, namely, the development of Americanism or American ethnic identity (Gleason 1980, 56–57). All the criteria for recognizing homeland space are met, including the stiffest test—the evidence of widespread ethnogenesis (the making of Americans), measured through historically high rates of ethnic acculturation, assimilation, and ongoing multicultural interaction (Barkan 1995). Here, beyond the legitimate concerns to differentiate "dominant" and "minority" cultures in the United States and the relations between them, lies the most remarkable feat of asserting a collective identity in the face of great diversity, a record no less of cultural development, self-determination, resistance, contest, persistence, retention, and frequently vehement attachment to place than any case within its borders (Zelinsky 1990). Fortunately, a penetrating geographical examination of the political and cultural dimensions of this homeland has appeared within recent years (Zelinsky 1988).

American Indian Homelands. American Indian peoples residing within the United States have a presumptive right to maintain traditional ways of life if they choose to (about 40 Indian cultures became extinct after European contact and lost the chance). The official option is by living on a reservation, and not all Indians choose reservation life. Only a few reservations now are located on territory historically identified with their groups (for example, the Navajo and Kiowa), and even fewer represent anything close to the amount of territory once regarded as homelands. Notwithstanding shifting Indian homelands prior to American interference, resulting from pressures on resources, culture change, and warfare, systematic dispossession at the hands of Euro-Americans produced a deeply fragmented modern pattern of arbitrarily imposed reservations, often in locations far removed from ancestral lands. Without addressing issues of the size and resource base of reservations in relation to Indian demography and survival, not to mention issues of government treatment and culture maintenance (for an excellent overview, see Frantz 1999), there is no question about recognizing Indian reservations de jure as a special type of homeland in America.

Given the extreme differences in size and physical character of the reservations, however, as well as the size and dynamics of the groups associated with them, it is problematic which reservations actually function de facto as homelands in the terms discussed above. Indians have citizenship in the United States, to which they owe allegiance, and in individual Indian nations or tribes. Many reservations are run by tribal governments with substantial autonomy, others by different means. Within the family of Indian communities some groups are culturally vigorous, economically self-confident, and politically astute; other groups show interest in various forms of pan-Indian cooperation and unified development, while a few interests seek secession from the United States. Suffice it to say that much work of a geographical nature can be done to clarify further which Indian communities (among the 109 distinct Indian cultures recognized by anthropologists, or the 298 reservations, or the 307 distinct groups recognized by the United States government) occupy and maintain homelands in the sense discussed here (O'Leary and Levinson 1991, xxiv; Prucha 1990). Because a systematic classification of Indian homelands is beyond precise definition at the moment, a crude approximation of the geographical dimensions of the problem can be gained by associating the present distribution of reservations with the distribution of Indian land areas judicially established (map 15.1).

Other Homelands. Among people of European ancestry, the candidate list of makers of homelands in the United States is very short indeed. Of all the possibilities suggested, the Hispano homeland seems the most robust (map 15.1). Alvar Carlson and Nostrand have demonstrated in detailed studies of the Hispano area centered in New Mexico that a hybrid people developed there over a period of four hundred years in decided geographical isolation, with a distinct sense of identity, able to occupy and control land exclusively in the form of a coherent spatial entity (Carlson 1990; Nostrand 1992). Clear evidence exists that the group impressed its culture upon the landscape and also that members prefer to live in the area and for self-imposed "exiles" to return. Whatever resistance the group put up to Anglo intrusion in the nineteenth century was easily overcome (Meinig 1971, 35) and any residual sense of peoplehood is today largely contained within a broader allegiance to the United States. Nostrand stresses that the Hispano homeland has been in obvious decline since 1900, and that its significance seems fainter with the passing years and in light of the burgeoning modern Mexican American presence in the borderland region as a whole. Today, it is becoming difficult to distinguish the old Hispano homeland within what I would term a broader Mexican-dominated Latino ethnic substrate that occupies the whole southwestern rim of the United States.

The Cajun people have also created a demonstrable homeland in southern Louisiana (map 15.1). Malcolm Comeaux (1977) has drawn attention to their unique adaptation to a difficult environment bypassed by others and stressed their complex population origins, despite the salience of French language use (Comeaux 1996a, 1996b). Eric Waddell (1983) examined the group's cultural distance from Francophone culture in Quebec, and Estaville (1986, 1988, 1993) has documented the group's occupation of territory over time. There seems little doubt that the Cajuns evolved ethnically in place and became thoroughly indigenous over the course of nearly 250 years. The homeland has long been split in two units by the Atchafalaya Basin Swamp but has been in decline with the intrusions of modernity. While the culture has returned the infiltration to some extent, with distinctive music and food preferences that have achieved wider popularity, it is hard to see evidence of any movement for ethnic self-determination in the political sphere—cooptation by the Anglo establishment when useful, but no serious separatist sentiments. As with the Hispanos in the Southwest, the Cajuns have seemingly welcomed the general benefits of American culture, and maintained distinctiveness largely as a legacy of relative isolation than of any rebellion against the dominant political and cultural regime.

This is not the place to examine the extent to which the Mormons might be considered in some way an American ethnic group, but the history of religious persecution, expulsion and relocation, together with the extensive set of singular social practices and environment adjustments they made in Utah set them apart in the mid-nineteenth century from mainstream American society. As Donald Meinig (1965) showed long ago, the Mormons created a sharply delineated culture region in the American West that has endured. If it did not develop in as prolonged an isolation as that of the Hispanos, the isolation was just as emphatic, and the conscious desire to be different from other social groups set the group on a different plane. The theocratic society of early Mormondom offers a rare illustration of political-cultural separatist impulses in the American context. The group petitioned Congress to recognize the state of Deseret in July 1849. Although it was never sanctioned, it did serve "as the sole functioning civil government" for the Intermontane Basin from then until February 1851 (May 1994, 90). Thus, a sense of separate group identity drawn from several sources, immediate declaration of a "home" in Utah with spatial coherence, control of territory, initial exclusivity, political institutions, defense of religious convictions, considerable debate over independence from United States government authority, and landscape impress, all make a strong case for at least a fleeting homeland (Bennion 1980; Francaviglia 1978). There are some similarities here with European-style protonationalism of the same period, without the long prior ethnic gestation. Subsequent religious and political developments among Mormons kept the group within the American family of communities, and with a universalizing missionary goal the church has sought to extend its influence far beyond its Utah stronghold. As the non-Mormon presence has grown in that state, and as Mormon settlement has expanded throughout the West, it might be appropriate to regard this case as one that made a historic transition from a strong homeland initiative to a modern regional cultural substrate.

Culture Regions

When too few of the criteria for homelands are met by American groups in regional space, they may more plausibly be seen as defining or helping to define culture regions and subregions within the national homeland. Zelinsky's famous "contemporary map of American culture areas" (1973, 118–19), which he saw no reason to alter radically for the revised edition of the work (Zelinsky 1992), shows several regions of "uncertain status," all in the southern portion, as well as large zones that invite closer investigation. Key to his scheme is that most of the regions he defined have drawn on several sources of culture, with New England being the sole exception. These culture areas have, for the most part, then, developed from various exogenous sources, and there was over time too much migration, interaction, and cultural change to have engendered in these areas a clearly separate people, with their own cultural goals and political strategies, independent of or even at odds with those of their neighbors. Hence, there is no reason to see the English settlers of New England or the more diverse inhabitants of Pennsylvania (including the Amish) as new and sharply distinct breeds of people bent on contrasting and incompatible national purposes. In the seventeenth and eighteenth centuries, before national independence, that is, their homelands clearly lay in Britain, Holland, the German states, and other European venues, and after the American Revolution if there was a new homeland aborning it was, however tenuously at first, a United States of America. There was no such deep isolation and neglect by a distant government as occurred with the Hispanos of the upper Río Grande in New Mexico.

Likewise, the Upper South and Texas hardly produced new, aggressively self-conscious peoples outside the routine pattern of American frontier extension (Mitchell 1991; Meinig 1969; Jordan and Kaups 1989). The Upper South is, in fact, clearly designated as culture area "III-c" on Zelinsky's map. Anglo-Texans encountered and reacted to alien cultures in principle little differently from those farther north in the frontier zone. Robert D. Mitchell (1978) has offered a schematic map of secondary cultural staging areas beyond the Appalachians, and it might be useful to regard Texas more as a particularly elaborate case of a major western staging area for the reformulation of cultural impulses being carried further west than as a curious homeland for people who were not that different from their Tennessee or Alabama fathers and mothers. This logic is not to disregard the complicating encounter with Mexico nor the experience of the South's secession (in which Texas played as much a contributing role to the Confederate project as it did later to national integration).

Ethnic Substrates and Enclaves

The concept of ethnic substrates emerged in reaction to the preoccupation with mapping and analyzing ethnic majorities across areas. While having a numerical majority or a plurality within a certain territory might well say something about an ethnic group, it does not follow that the lack of a majority or plurality dooms groups to abject dependence on the will of the leading group. Hence, an ethnocultural substrate can be defined as: "A zone within which a particular ethno-cultural group is consistently above a certain minimum proportion of the total population, thereby constituting a recurrent presence, even if a minority, from locality to locality within the zone, which may influence the broad community values, regional identity, and landscape character of the zone as a whole" (Conzen 1993, 19–21). This ethnocultural latency may be particularly significant in forging "regimes of cooperation" among interest groups across the social spectrum in seeking to achieve particular goals—temperance, railroad regulation, opposition to abortion, and bilingualism come to mind as examples.

Though the concept has been little applied so far (but see Helzer 1998, and Jordan-Bychkov's chap. 5), there is no reason why the distribution and sociopolitical significance of African Americans across the Deep South, Creoles along the coast, and Tejanos in southern Texas, might not be interpreted in terms of ethnic and ethnoracial substrates rather than as homelands (map 15.2). The idea that homelands are for sharing is antithetical to the basic notion of a more or less exclusive territory that one indigenous group controls to advance its own cultural goals. White southerners uninterested in racial equality might have assumed that at one time the South could be run for their exclusive benefit, but their dependence on the other resident group renders the term difficult to accept for two groups with such divergent goals. The case of the Tejanos, with the proximity of Mexico, the question of what proportion of the group has long Texas ancestry, and the issue of how culturally different the Tejanos are from the Mexicanos, could well be more suitably examined from the perspective of an ethnic substrate in South Texas. Daniel D. Arreola (1993, 61–62) has pointed out that the Texas-Mexican area of South Texas cannot be regarded a homeland "in the traditional rural sense," but rather a "regional ethnic enclave."

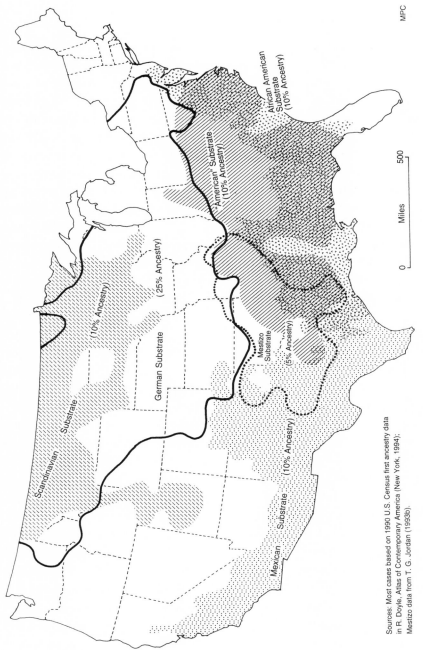

Sources: Most cases based on 1990 U.S. Census first ancestry data in R. Doyle, Atlas of Contemporary America (New York, 1994); Mestizo data from T. G. Jordan (1993b).

Map 15.2. Selected ethnocultural substrates in the United States.

Ethnocultural Islands and Archipelagos

The most venerable concept of space relating to cultural groups is the concept of "ethnic islands" (Raitz 1978). It came into use to describe plainly visible ethnic clustering at limited scales and gained currency because so many instances can be documented around the United States and elsewhere. Sometimes, these ethnic islands acquire apt or ironic labels, such as the concentration of German, Luxembourg, and Polish Catholics in central Minnesota, dubbed "The Minnesota Holy Land" in the literature (Vogeler 1976; Dockendorff 1986). It is unlikely that ethnic islands are going to be considered ethnic homelands, given their small size and especially if they represent a minute proportion of the group's population. An archipelago of ethnic islands also does not make up a homeland (Conzen 1990, 242–45; 1993, 24; 1996). Such a situation is well illustrated among others by the Amish, who set down some roots in a "hearth area" in southeast Pennsylvania two centuries ago, but whose development in the United States has been predicated on dispersion, not occupying large contiguous areas (map 15.3; Crowley 1978).

Closing Observations

The essential argument of this review can be summarized as follows. First, homelands, as cases of geographical culture regions with special character (of long ancestry, once exclusively occupied, and now or once politically independent), result from self-consciously separate "peoples" whose "sense of peoplehood" and control of territory sustains the fact of, or ambitions for, political and cultural independence. Second, in an American context, the biggest and most successful homeland is the American national homeland, developed over the last two hundred years. Prior American Indian homelands were extinguished, relocated, or massively reduced by the United States government by treaty and force, but many survive as official entities in the form of reservations. Third, historic regional homelands originated through settlement and ethnogenesis only in locations beyond American sovereignty, and once hegemony from coast to coast was established, no new subnational homelands emerged, and old ones atrophied. And fourth, with the development of national American society, regional populations (both folk and modern) and immigrant groups have occupied parts of the country in varied proportions and

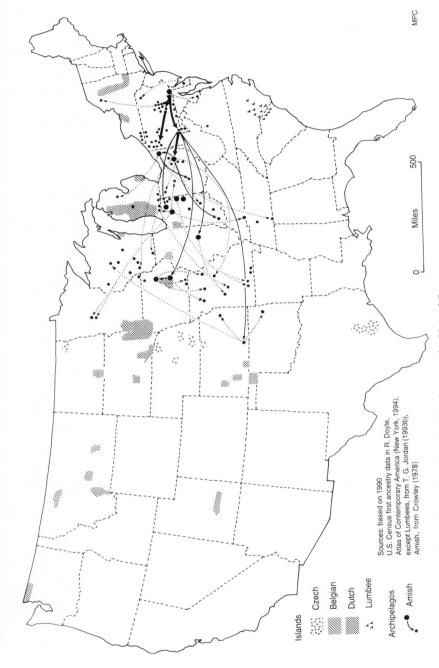

Islands

Czech

Belgian

Dutch

Lumbee

Archipelagos

Amish

Sources: based on 1990
U.S. Census first ancestry data in R. Doyle,
Atlas of Contemporary America (New York, 1994),
except Lumbees, from T. G. Jordan (1993b),
Amish, from Crowley (1978)

0   500

Miles

Map 15.3. Selected ethnocultural islands and archipelagos in the United States.

degrees of intermixture, leading to the spatial definition of a number of composite American culture regions, complete with "hearths," diffusion paths, recombinative staging areas, and heartlands. Because of cultural contact and admixture, high rates of migration and mutual cultural influence, these particular culture regions are not homelands for any one group within the collective array.

In the interest of conceptual development, a set of criteria for recognizing homelands can be set up along three dimensions: cultural identity (ethnogenesis, indigenization over time, and geographical isolation), territoriality (control over land and resources, political institutions, spatial coherence), and loyalty to place (defense of the homeland, compulsion to live there, and production of nationalistic landscape symbols).

The conceptual significance of homelands in the cultural and ethnic regionalism of America is twofold. First of all, the term *homeland*, applied at the subnational scale, serves to distinguish groups and areas with a long, ancestral, once largely exclusive, and distinctive mutual association from more diffuse, multicultural associations and areas. The fate of ethnic homelands in America, however, is that they have mostly faded as strong culturally organized expressions of identity, while ethnicity as such, which is not so place-bound, is flourishing—now a new kind of bargaining chip in the postmodern age.

This creates a second basis for significance, that "homeland" might become a geographical dimension in issues of ethnic and other group "rights" in American life, which promise to become even more contentious than before (Tesón 1998). Eugeen Roosens observed that "the ethnic unit is one of the few organizational forms that, on the macro-level, offers stability in a time of decline of authority in all its forms. . . . The nation, too, has lost much of its ideological foundation and power of attraction because there are only a few countries left which emphasize a truly national ideology and because there have been no wars between the nations of the West since World War II" (Roosens 1989, 17). Hence, there are potential future political uses of the idea of homeland in American multicultural discourse that, if they appear, should be examined for their motives and potential uses and misuses of history and geography.

Elsewhere the history of the homeland concept has been political. *Homeland* has been invested with emotive force and used in struggles for ethnic political power. Therefore, it is prudent to ask what are the contemporary motives underlying homeland recognition in the United States? Is the drive to recognize numerous homelands there simply to serve disinterested geographical curiosity or to serve the objectives of "pressure groups with a noble face," because culture is "a useful weapon in peaceful settings" (Roosens 1989, 14)? If the former, what features of a regional population occupying a particular segment of United States territory suggest a homeland rather than simply a geographically "patterned differentiation" (Petersen 1980, 234) or simply a culture area (which, in any case, is rarely simple)? What larger political purposes are served by recognizing ethnic and cultural homelands within the nation? Does academic recognition of homelands today lay the groundwork for special status tomorrow? Further examination of the homeland concept is without doubt desirable—particularly of American Indian homelands, because they are, above all, an official legacy of a harrowing chapter in American cultural and diplomatic history, and they are as keen a challenge to understanding the geographical dimensions of social justice as can be presented by any group in the American nation.

Aside from the remnant homelands of Native Americans officially recognized as territorial reservations, and a couple of other historical cases of waning significance, most other subnational cultural groups today occupy geographical space in ways better articulated by such concepts as ethnic islands, ethnic substrates, and cultural regions.

REFERENCES

Introduction

Billington, Ray Allen. 1973. *Frederick Jackson Turner: Historian, Scholar, Teacher.* New York: Oxford Univ. Press.

Carlson, Alvar W. 1971. "The Rio Arriba: A Geographic Appraisal of the Spanish-American Homeland (Upper Rio Grande Valley, New Mexico)." Ph.D. diss., Univ. of Minnesota.

Carlson, Alvar W. 1990. *The Spanish-American Homeland: Four Centuries in New Mexico's Río Arriba.* Baltimore: Johns Hopkins Univ. Press.

Conzen, Michael P. 1993. "Culture Regions, Homelands, and Ethnic Archipelagos in the United States: Methodological Considerations." *Journal of Cultural Geography* 13, no. 2: 13–29.

Nostrand, Richard L. 1992. *The Hispano Homeland.* Norman: Univ. of Oklahoma Press.

Turner, Frederick Jackson. 1920. "The Significance of the Frontier in American History." *The Frontier in American History,* 1–38. New York: Henry Holt.

Webb, Walter Prescott. 1931. *The Great Plains.* Boston: Ginn.

Webb, Walter Prescott. 1957. "The American West, Perpetual Mirage." *Harper's Magazine* 214, no. 1284: 25–31.

Wishart, David J. 1987. "Settling the Great Plains, 1850–1930: Prospects and Problems." In *North America: The Historical Geography of a Changing Continent,* ed. Robert D. Mitchell and Paul A. Groves, 255–78. Totowa, N.J.: Rowman & Littlefield.

CHAPTER 1: *The New England Yankee Homeland*

Ackerman, Edward. 1941. "Sequent Occupance of a Boston Suburban Community." *Economic Geography* 17: 61–74.

Allen, David G. 1982a. *In English Ways.* New York: Norton.

Allen, David G. 1982b. *"Vacuum Domicilium:* Social and Cultural Landscape of Seventeenth Century New England." In *New England Begins: The Seventeenth Century,* 1: 1–52. Boston: Museum of Fine Arts.

Bailyn, Bernard. 1964. *The New England Merchants in the Seventeenth Century.* New York: Harper & Row.

Banks, Charles E. 1937. *Topographical Dictionary of 2885 English Emigrants to New England,* ed. E. E. Brownell. Philadelphia: Genealogical Publishing.

Bearse, Ray, ed. 1971. *Massachusetts*. 2d ed. Boston: Houghton Mifflin.

Bowden, Martyn J. 1992a. "Invented Tradition and Academic Convention in Geographical Thought about New England." *GeoJournal* 26, no. 2: 187–94.

Bowden, Martyn J. 1992b. "The Invention of American Tradition." *Journal of Historical Geography* 18, no. 1: 3–26.

Bowden, Martyn J. 1994a. "Culture and Place: English Sub-Cultural Regions in New England in the Seventeenth Century." *Connecticut History* 35, no. 1: 68–146.

Bowden, Martyn J. 1994b. "Place-Names and Cultural Geography: Defining New England's Cultural Regions in the Seventeenth and Eighteenth Centuries." Paper presented to Eastern Historical Geography Association, Halifax, Nova Scotia.

Brooke, John L. 1989. *The Heart of the Commonwealth*. New York: Cambridge Univ. Press.

Brown, R. J. 1979. *The English Country Cottage*. London: Robert Hale.

Brown, Richard D. 1978. *Massachusetts: A History*. New York: Norton.

Brunskill, R. W. 1971. *Illustrated Handbook of Vernacular Architecture*. London: Faber.

Bushman, Richard L. 1970. *From Puritan to Yankee: Character and Social Order in Connecticut, 1690–1765*. New York: Norton.

Bushman, Richard L. 1984. "American High-Style and Vernacular Cultures." In *Colonial British America*, ed. Jack P. Greene. Baltimore: Johns Hopkins Univ. Press.

Candee, Richard M. 1976a. "The Architecture of Maine's Settlement: Vernacular Architecture to About 1720." In *Maine Forms of American Architecture*, ed. Deborah Thompson, 15–44. Camden: Downeast.

Candee, Richard M. 1976b. "Wooden Buildings in Early Maine and New Hampshire: A Technological and Cultural History, 1600–1720." Ph.D. diss., Univ. of Pennsylvania.

Candee, Richard M. 1992. *Building Portsmouth*. Portsmouth: Portsmouth Advocates.

Clark, Charles E. 1990. *Maine: A History*. Hanover: Univ. of New England Press.

Clifton-Taylor, Alec. 1962. *The Pattern of English Building*. London: Batsford.

Connally, Ernest A. 1960. "The Cape Cod House: An Introductory Study." *Journal of the Society of Architectural Historians* 19, no. 2: 47–56.

Cummings, Abbott L. 1967. "The Garrison House Myth." *Historical New Hampshire* 22, no. 1: 3–17.

Cummings, Abbott L. 1979. *The Framed Houses of Massachusetts Bay, 1625–1725*. Cambridge: Harvard Univ. Press.

Cummings, Abbott L. 1994. "Connecticut and Its Building Traditions." *Connecticut History* 35, no. 1: 192–233.

Daniels, Bruce C. 1983. *Dissent and Conformity on Narragansett Bay*. Middletown: Wesleyan Univ. Press.

Deetz, James. 1979. "Plymouth Colony Architecture: Archaeological Evidence from the Seventeenth Century." In *Architecture in Colonial Massachusetts*, ed. Abbott L. Cummings, 43–60. Boston: Colonial Society of Massachusetts.

Demos, John. 1971. *A Little Commonwealth*. New York: Oxford Univ. Press.

Foster, George M. 1960. *Culture and Conquest: America's Spanish Heritage*. Chicago: Quadrangle Books.

Garrison, J. Ritchie. 1991. *Landscape and Material Life in Franklin County, Massachusetts, 1770–1860*. Knoxville: Univ. of Tennessee Press.

Glassie, Henry. 1968. *Pattern in the Material Folk Culture of the Eastern United States*. Philadelphia: Univ. of Pennsylvania Press.

Greene, Jack P. 1988. *Pursuits of Happiness*. Chapel Hill: Univ. of North Carolina Press.

Hansen, Marcus. 1939. "The Settlement of New England." In *Handbook of the Linguistic Geography of New England*, ed. Hans Kurath, 61–104. Providence: Brown Univ. Press.

Johnson, Edward. 1654. *Wonder-working Providence of Sion's Saviour in New England*, ed. J. Franklin Jameson. Reprint, 1910. New York: Scribner's.

Kelly, J. Frederick. 1963. *Early Domestic Architecture of Connecticut*. 1924. Reprint. New York: Dover.

Lancaster, Clay. 1972. *The Architecture of Historic Nantucket*. New York: McGraw-Hill.

Lancaster, Clay. 1979. *Nantucket in the Nineteenth Century*. New York: Dover.

Lockridge, Kenneth A. 1970. *A New England Town: The First Hundred Years*. New York: Norton.

Mencken, H. L. 1936. *The American Language*. 4th Edition. New York: Knopf.

Miller, Amelia. 1983. *Connecticut River Doorways*. Boston: Boston Univ.

Morison, Samuel E. 1960. *The Story of the "Old Colony" of New Plymouth*. New York: Knopf.

Postgate, Malcolm R. 1973. "Field Systems of East Anglia." In *Studies in Field Systems in the British Isles*, ed. Alan R. H. Baker and Robin A. Butlin, 281–324. Cambridge: Cambridge Univ. Press.

Powell, Sumner C. 1965. *Puritan Village: The Formation of a New England Town*. New York: Doubleday.

Price, Jacob M. 1974. "Economic Function and the Growth of American Port Towns in the Eighteenth Century." *Perspectives in American History* 8: 123–85.

Roxby, Percy M. 1928. "East Anglia." In *Great Britain: Essays in Regional Geography*, ed. Alan G. Ogilvie, 143–66. Cambridge: Cambridge Univ. Press.

Rutman, Darret B. 1965. *Winthrop's Boston: A Portrait of a Puritan Town, 1630–1649*. Chapel Hill: Univ. of North Carolina Press.

Sauer, Carl O. 1941. "Settlement of the Humid East." In *Climate and Man*, 157–66. Yearbook of the U.S. Department of Agriculture. Washington, D.C.: Government Printing Office.

Smith, Chard P. 1946. *The Housatonic: Puritan River*. New York: Rinehart.

St. George, Robert B. 1982. "'Set Thine House in Order': Domestication of the Yeomanry in Seventeenth-Century New England." In *New England Begins: The Seventeenth Century*, ed. Jonathan L. Fairbanks and Robert F. Trent, 2: 159–83. Boston: Museum of Fine Arts.

St. George, Robert B. 1985. "Artifacts of Regional Consciousness in the Connecticut

River Valley, 1700–1780." In *The Great River: Art and Society of the Connecticut Valley, 1635–1820*, ed. William N. Hosley, Jr., and Gerald W. R. Ward, 29–40. Hartford: Wadsworth Atheneum. Reprinted in *Material Life in America, 1600–1860*, ed. Robert Blair St. George, 335–56. Boston: Northeastern Univ. Press, 1988.

Steinitz, Michael. 1986. "Rethinking Geographical Approaches to the Common House: The Evidence from Eighteenth-Century Massachusetts." In *Perspectives in Vernacular Architecture*, vol. 3, ed. Thomas Carter and Bernard J. Herman, 16–26. Columbia: Univ. of Missouri Press.

Thirsk, Joan. 1967. "The Farming Regions of England." In *The Agrarian History of England and Wales*, vol. 4: *1500–1640*, ed. Joan Thirsk, 1–112. Cambridge: Cambridge Univ. Press.

Upton, Dell. 1979. "Architectural Change in Colonial Rhode Island: The Mott House as a Case Study." *Old-Time New England* 69, no. 1: 18–33.

Van Deventer, David E. 1976. *The Emergence of Provincial New Hampshire, 1623–1741*. Baltimore: Johns Hopkins Univ. Press.

Wertenbaker, Thomas J. 1947. *The Puritan Oligarchy: The Founding of American Civilization*. New York: Grosset & Dunlap.

Whittlesey, Derwent. 1929. "Sequent Occupance." *Annals of the Association of American Geographers* 19: 162–65.

Wood, Joseph S. 1978. "The Origin of the New England Village." Ph.D. diss., Pennsylvania State Univ.

Wood, Joseph S. 1986. "The New England Village as an American Vernacular Form." In *Perspectives in Vernacular Architecture*, vol. 2, ed. Camille Wells, 54–63. Columbia: Univ. of Missouri Press.

Wood, Joseph S. 1997. *The New England Village*. Baltimore: Johns Hopkins Univ. Press.

Wright, John K. 1934. "Regions and Landscapes of New England." In *New England's Prospect: 1933*, ed. John K. Wright, 14–49. New York: American Geographical Society.

CHAPTER 2: *The Pennsylvanian Homeland*

Barley, M. W. 1961. *The English Farmhouse and Cottage*. London: Routlege & Kegan Paul.

Benson, Adolph B. 1966. *Peter Kalm's Travels in North America*. New York: Dover Publications.

Clifton, Ronald. 1971. "Forms and Patterns: Room Specialization in Maryland, Massachusetts, and Pennsylvania, Family Dwellings, 1725–1834." Ph.D. diss., Univ. of Pennsylvania.

Dornbusch, Charles H., and John K. Heyl. 1958. *Pennsylvania German Barns*. Pennsylvania German Folklore Society, no. 21. Allentown, Pa.: Schlechter's.

Ensminger, Robert F. 1992. *The Pennsylvania Barn: Its Origin, Evolution, and Distribution in North America*. Baltimore: Johns Hopkins Univ. Press.

Garber, John P. 1934. *The Valley of the Delaware and Its Place in American History*. Reissued, 1969. Port Washington, N.Y.: Ira J. Friedman.

Glass, Joseph. 1971. "The Pennsylvania Culture Region: A Geographical Interpretation of Barns and Farmhouses." Ph.D. diss., Pennsylvania State Univ.

Glass, Joseph. 1986. *The Pennsylvania Culture Region: A View from the Barn*. Ann Arbor: UMI Research Press.

Kniffen, Fred. 1965. "Folk Housing: Key to Diffusion." *Annals of the Association of American Geographers* 55: 549–77.

Lemon, James. 1972. *The Best Poor Man's Country*. Baltimore: Johns Hopkins Univ. Press.

Lewis, Peirce F. 1972. "Small Town in Pennsylvania." *Annals of the Association of American Geographers* 62: 323–51.

Marshe, W. 1801. "Journal." *Collections of the Massachusetts Historical Society*, 1st ser., 7.

McDonald, Forrest, and Ellen McDonald. 1980. "The Ethnic Origins of the American People, 1790." *William and Mary Quarterly* 37, no. 2: 179–99.

Mittelberger, Gottlieb. 1960. *Journey to Pennsylvania*, ed. and trans. Oscar Handlin and John Clive. Cambridge, Mass.: Belknap Press of Harvard Univ. Press.

Myers, Albert. 1912. *Narratives of Early Pennsylvania and West New Jersey and Delaware, 1630–1707*. New York: Scribner's.

Pillsbury, Richard. 1968. "The Urban Street Patterns of Pennsylvania before 1815: A Study in Cultural Geography." Ph.D. diss., Pennsylvania State Univ.

Pillsbury, Richard. 1977. "Patterns in the Folk and Vernacular House Forms of the Pennsylvania Culture Region." *Pioneer America* 9, no. 3: 12–31.

Pillsbury, Richard. 1987. "The Pennsylvania Culture Area: A Reappraisal." *North American Culture* 3, no. 2: 37–54.

Purvis, Thomas L. 1987. "Patterns of Ethnic Settlement in Late Eighteenth-Century Pennsylvania." *Western Pennsylvania Historical Magazine* 70, no. 1: 107–22.

Weaver, William. 1982. *A Quaker Woman's Cookbook: The Domestic Cookery of Elizabeth Ellicott Lea*. Philadelphia: Univ. of Pennsylvania Press.

Zelinsky, Wilbur. 1977. "The Pennsylvania Town: An Overdue Geographical Account." *Geographical Review* 67, no. 2: 127–47.

CHAPTER 3: *Old Order Amish Homelands*

Crowley, William K. 1978. "Old Order Amish Settlement: Diffusion and Growth." *Annals of the Association of American Geographers* 68, no. 2: 249–64.

Hostetler, John A. 1981. *Amish Life*. Kitchener: Herald.

Hostetler, John A. 1993. *Amish Society*. Baltimore: Johns Hopkins Univ. Press.

Kraybill, Donald B. 1989. *The Riddle of Amish Culture*. Baltimore: Johns Hopkins Univ. Press.

Lamme, Ary J., III, and Douglas B. McDonald. 1984. "Recent Amish Settlement in the North Country." *Material Culture* 16, no. 2: 77–91.

Lamme, Ary J., III, and Douglas B. McDonald. 1993. "The North Country Amish Homeland." *Journal of Cultural Geography* 13, no. 2: 107–18.

Wittmer, Joe. 1990. *The Gentle People: Personal Reflections of Amish Life.* Minneapolis: Educational Media Corporation.

CHAPTER 4: *Blacks in the Plantation South*

Aiken, Charles S. 1985. "New Settlement Patterns of Blacks in the Plantation South." *Geographical Review* 75, no. 4: 283–404.

Aiken, Charles S. 1987. "Race as a Factor in Municipal Underbounding." *Annals of the Association of American Geographers* 77, no. 4: 564–79.

Aiken, Charles S. 1990. "A New Type of Black Ghetto in the Plantation South." *Annals of the Association of American Geographers* 80, no. 2: 223–46.

Aiken, Charles S. 1998. *The Cotton Plantation South since the Civil War.* Baltimore: Johns Hopkins Univ. Press.

Auchmutey, Jim. 1993. "Keepers of the Confederacy." *Atlanta Journal/Atlanta Constitution,* 19 January, B1 and B3.

Ball, Edward. 1998. *Slaves in the Family.* New York: Farrar, Straus & Giroux.

Brooke, James. 1998. "Blacks in U.S. Heartland Put New Focus on Africa." *New York Times,* 22 March, 6.

"Civil Rights Memorial Shadowed by Rebel Flag." 1989. *Atlanta Journal/Atlanta Constitution,* 5 November, A9.

Cobb, James C. 1996. "Community and Identity: Redefining Southern Culture." *Georgia Review* 50, no. 1: 9–24.

Cohen, William. 1991. *At Freedom's Edge: Black Mobility and the Southern White Quest for Racial Control, 1861–1915.* Baton Rouge: Louisiana State Univ. Press.

Crockett, Norman L. 1979. *The Black Towns.* Lawrence: Regents Press of Kansas.

Du Bois, W. E. Burghardt. 1903. *The Souls of Black Folk: Essays and Sketches.* Chicago: A. C. McClurg.

Du Bois, W. E. Burghardt. 1947. "Behold the Land." *New Masses,* 14 January, 18–20.

Gomez, Michael. 1998. *Exchanging Our Country Marks: The Transformation of African Identities in the Colonial and Antebellum South.* Chapel Hill: Univ. of North Carolina Press.

Halberstam, David. 1993. *The Fifties.* New York: Villard Books.

Haley, Alex. *Roots.* 1976. Garden City, N.Y.: Doubleday.

Hunter-Gault, Charlayne. 1992. *In My Place.* New York: Farrar, Straus & Giroux.

Johnson, Daniel M., and Rex R. Campbell. 1981. *Black Migration in America: A Social Demographic History.* Durham, N.C.: Duke Univ. Press.

Johnson, Rheta Grimsley. 1996. "Walton Dismisses a 'Mississippi' Much Changed." *Atlanta Journal/Atlanta Constitution,* 21 April, L10.

Mandle, Jay R. 1978. *The Roots of Black Poverty: The Southern Plantation Economy after the Civil War.* Durham, N.C.: Duke Univ. Press.

McAlister, Durwood. 1990. "Era of the Old South Is Gone, but Its Echoes Remain." *Atlanta Journal/Atlanta Constitution*, 21 October, D5.

Phillips, Michael M. 1998. "Some Blacks Are Torn by Trade Bill." *Wall Street Journal*, 11 March, A2.

Prunty, Merle C., and Charles S. Aiken. 1972. "The Demise of the Piedmont Cotton Region." *Annals of the Association of American Geographers* 62, no. 2: 283–306.

Redkey, Edwin S. 1969. *Black Exodus: Black Nationalist and Back-to-Africa Movements, 1890–1910*. New Haven: Yale Univ. Press.

Reed, Adolph, Jr. 1997. *W. E. B. Du Bois and American Political Thought: Fabianism and the Color Line*. New York: Oxford Univ. Press.

Sinclair, Ward. 1986. "Black Farmers: A Dying Minority." *Washington Post*, 18 February, A1 and A4.

Stack, Carol. 1996. *Call to Home: African Americans Reclaim the Rural South*. New York: Basic Books.

Taulbert, Clifton L. 1989. *Once upon a Time When We Were Colored*. Tulsa, Okla.: Council Oak Books.

U.S. Bureau of the Census. 1947. *United States Census of Agriculture, 1945: Color and Tenure of Farm Operator*. Vol. 2, chap. 3. Washington, D.C.: Government Printing Office.

U.S. Bureau of the Census. 1963. *1960 Census of Housing*. Vol. 1, pts. 2, 3, 5. Washington, D.C.: Government Printing Office.

U.S. Bureau of the Census. 1992. *1990 Census of Population, General Population Characteristics, United States*. 1990 CP-1-1. Washington, D.C.: Government Printing Office.

Walton, Anthony. 1996. *Mississippi: An American Journey*. New York: Knopf.

Woodward, C. Vann. 1955. *The Strange Career of Jim Crow*. 3d ed., 1974. New York: Oxford Univ. Press.

Woodward, C. Vann. 1960. *The Burden of Southern History*. Baton Rouge: Louisiana State Univ. Press.

Wright, Richard, and Edwin Rosskam. 1941. *Twelve Million Black Voices: A Folk History of the United States*. New York: Viking Press.

CHAPTER 5: *The Creole Coast*

Andrews, Jean. 1993. *Red Hot Peppers*. New York: Macmillan.

Andrews, Jean. 1995. "A Botanical Mystery: The Elusive Trail of the Datil Pepper to St. Augustine." *Florida Historical Quarterly* 74, no. 2: 132–47.

Atwood, E. Bagby. 1962. *The Regional Vocabulary of Texas*. Austin: Univ. of Texas Press.

Babington, Mima, and E. Bagby Atwood. 1961. "Lexical Usage in Southern Louisiana." *Publications of the American Dialect Society* 36: 1–24.

Bartram, William. 1791. *The Travels of William Bartram*. Reprint, 1958. 2 vols. New Haven: Yale Univ. Press.

Beckles, Hilary M. 1989. *White Servitude and Black Slavery in Barbados, 1627–1715.* Knoxville: Univ. of Tennessee Press.

Carney, Judith A. 1993. "From Hands to Tutors: African Expertise in the South Carolina Rice Economy." *Agricultural History* 67, no. 3: 1–30.

Cassidy, Frederic G., and R. B. LePage. 1980. *Dictionary of Jamaican English.* 2d ed. Cambridge: Cambridge Univ. Press.

Crawford, James M. 1978. *The Mobilian Trade Language.* Knoxville: Univ. of Tennessee Press.

Dillard, J. L. 1985. *Toward a Social History of American English.* Berlin: Mouton.

Dillard, J. L. 1987. "The Maritime (Perhaps Lingua Franca) Relations of a Special Variety of the Gulf Corridor." *Journal of Pidgin and Creole Languages* 2, no. 2: 244–49.

Dillard, J. L. 1992. *History of American English.* London: Longman.

Edwards, Jay. 1989. "The Complex Origins of the American Domestic Piazza-Veranda-Gallery." *Material Culture* 21, no. 2: 3–58.

Ekman, Sven. 1953. *Zoogeography of the Sea.* London: Sidgwick & Jackson.

Fagg, Daniel, Jr. 1989. "Raised Courthouses in North Carolina." *Material Culture* 21, no. 3: 1–8.

Flint, Timothy. 1828. *A Condensed Geography and History of the Western States.* 2 vols. Cincinnati: E. H. Flint.

Gritzner, Janet B. 1978. "Tabby in the Coastal Southeast." Ph.D. diss., Louisiana State Univ.

Hammond, Edwin H. 1964. "Classes of Land-Surface Form in the Forty-Eight States, U.S.A." *Annals of the Association of American Geographers* 54, no. 1: folded map supplement.

Harper, Francis, and Delma E. Presley. 1981. *Okefinokee Album.* Athens: Univ. of Georgia Press.

Jakle, John A., Robert W. Bastian, and Douglas K. Meyer. 1989. *Common Houses in American Small Towns.* Athens: Univ. of Georgia Press.

Jordan, Terry G. 1993. *North American Cattle-Ranching Frontiers.* Albuquerque: Univ. of New Mexico Press.

LeBon, J. W., Jr. 1971. "The Catahoula Hog Dog: A Folk Breed." *Pioneer America* 3, no. 2: 35–45.

Loftfield, Thomas C. 1993. "Caribbean Influence in the Seventeenth Century Landscape of Coastal North Carolina." *Pioneer America Society Transactions* 16: 72–73.

Manucy, Albert. 1992. *The Houses of St. Augustine, 1565–1821.* Gainesville: Univ. Press of Florida.

Mealor, W. Theodore, and Merle C. Prunty. 1976. "Open-Range Ranching in Southern Florida." *Annals of the Association of American Geographers* 66, no. 3: 360–76.

Morgan, Edmund S. 1975. *American Slavery, American Freedom: The Ordeal of Colonial Virginia.* New York: Norton.

Murphree, Nellie. 1952. "Concrete, Texas." In *The Handbook of Texas*, ed. Walter P. Webb and H. Bailey Carroll, 1: 391. Austin: Texas State Historical Association.

Olmsted, Frederick L. 1857. *A Journey through Texas.* Reprint, 1978. Austin: Univ. of Texas Press.

Oszuscik, Philippe. 1992a. "African-Americans in the American South." In *To Build in a New Land,* ed. Allen G. Noble, 157–76. Baltimore: Johns Hopkins Univ. Press.

Oszuscik, Philippe. 1992b. "Passage of the Gallery and Other Caribbean Elements from the French and Spanish to the British in the United States." *Pioneer America Society Transactions* 15: 1–14.

Oszuscik, Philippe. 1994. "Comparisons between Rural and Urban French Creole Housing." *Material Culture* 26, no. 3: 1–36.

Otto, John S. 1986. "The Origins of Cattle-Ranching in Colonial South Carolina, 1670–1715." *South Carolina Historical Magazine* 87, no. 1: 117–24.

Pike, Ruth. 1967. "Sevillian Society in the Sixteenth Century: Slaves and Freedmen." *Hispanic American Historical Review* 47, no. 3: 344–59.

Varner, John G., and Jeanette J. Varner. 1983. *Dogs of the Conquest.* Norman: Univ. of Oklahoma Press.

Vlach, John M. 1976. "The Shotgun House: An African Architectural Legacy." *Pioneer America* 8, no. 1–2: 36–70.

Waibel, Leo. 1941. "The Tropical Plantation System." *Scientific Monthly* 52, no. 2: 156–60.

CHAPTER 6: *Nouvelle Acadie*

Arceneaux, William. 1981. *Acadian General: Alfred Mouton and the Civil War.* 2d ed. Lafayette: Univ. of Southwestern Louisiana Center for Louisiana Studies.

Audisio, Gabriel. 1988. "Crisis in Baton Rouge, 1840–1860: Foreshadowing the Demise of Louisiana's French Language?" *Louisiana History* 29: 358–63.

Brandon, Elizabeth. 1972. "The Socio-Cultural Traits of the French Folksong in Louisiana." *Revue de Louisiane* 1: 19–39.

Brasseaux, Carl A. 1978. "Acadian Education: From Cultural Isolation to Mainstream America." In *The Cajuns: Essays on Their History and Culture,* ed. Glenn R. Conrad. Lafayette: Univ. of Southwestern Louisiana Center for Louisiana Studies.

Brasseaux, Carl A. 1987. *The Founding of New Acadia: The Beginnings of Acadian Life in Louisiana, 1765–1803.* Baton Rouge: Louisiana State Univ. Press.

Calhoun, James, ed. 1979. *Louisiana Almanac, 1979–1980.* Gretna, La.: Pelican Publishing.

Carleton, Eleanor B. 1948. "The Establishment of the Electric Telegraph in Louisiana and Mississippi." *Louisiana Historical Quarterly* 31: 3–68.

Case, Gladys C. 1973. *The Bayou Chene Story: A History of the Atchafalaya Basin and Its People.* Detroit: Harlo Press.

Clark, Andrew Hill. 1968. *Acadia: The Geography of Early Nova Scotia to 1760.* Madison: Univ. of Wisconsin Press.

Comeaux, Malcolm. 1972. *Atchafalaya Swamp Life: Settlement and Folk Occupations.* Baton Rouge: Louisiana State Univ. School of Geoscience.

Comeaux, Malcolm. 1978. "Louisiana Acadians: The Environmental Impact." In *The Cajuns: Essays on Their History and Culture*, ed. Glenn R. Conrad. Lafayette: Univ. of Southwestern Louisiana Press.

Comeaux, Malcolm. 1992. "Cajuns in Louisiana." In *To Build in a New Land: Ethnic Landscapes in North America*, ed. Allen G. Noble. Baltimore: Johns Hopkins Univ. Press.

Comeaux, Malcolm. 1996. "The Acadians of Canada and the Cajuns of Louisiana: Cultural Change over Time and Distance." In *Ethnic Persistence and Change in Europe and America*, ed. Klaus Frantz and Robert A. Sauder. Innsbruck: Univ. of Innsbruck.

Davis, Edwin A. 1965. *Louisiana, a Narrative History*. Baton Rouge: Claitor's Book Store.

Davis, Edwin A., ed. 1968. *The Rivers and Bayous of Louisiana*. Baton Rouge: Louisiana Education Research Association.

De Grummond, Jewel L. 1949. "A Social History of St. Mary Parish, 1845–1860." *Louisiana Historical Quarterly* 32: 17–102.

Edmonds, David C. 1979. *Yankee Autumn in Acadiana: The Great Texas Overland Expedition through Southwestern Louisiana (October-December 1863)*. Lafayette, La.: Acadiana Press.

Estaville, Lawrence E., Jr. 1984. "The Louisiana French Culture Region: Geographic Morphologies in the Nineteenth Century." Ph.D. diss., Univ. of Oklahoma.

Estaville, Lawrence E., Jr. 1986. "Mapping the Cajuns." *Southern Studies* 25: 163–71.

Estaville, Lawrence E., Jr. 1987. "Changeless Cajuns: Nineteenth-Century Reality or Myth?" *Louisiana History* 28: 117–40.

Estaville, Lawrence E., Jr. 1988. "Were the Nineteenth-Century Cajuns Geographically Isolated?" *Geoscience and Man* 25: 85–95.

Estaville, Lawrence E., Jr. 1989. *Confederate Neckties: Louisiana Railroads in the Civil War.* Ruston, La.: McGinty Publications, Louisiana Tech Univ.

Estaville, Lawrence E., Jr. 1990. "The Louisiana French Language in the Nineteenth Century." *Southeastern Geographer* 30: 105–20.

Estaville, Lawrence E., Jr. 1993. "The Louisiana-French Homeland." *Journal of Cultural Geography* 13: 31–45.

Foret, Michael J. 1980. "Aubry, Foucault, and the Attakapas Indians, 1765." *Attakapas Gazette* 15: 60–62.

Fortier, Alcée. 1891. "The Acadians of Louisiana and Their Dialect." *Modern Language Association of America Publications* 6: 64–94.

Ginn, Mildred K. 1940. "A History of Rice Production in Louisiana to 1896." *Louisiana Historical Quarterly* 23: 544–88.

Griffiths, Naomi. 1973. *The Acadians: Creation of a People*. New York: McGraw-Hill; Toronto: Ryerson.

Heck, Robert W. 1978. "Building Materials in the Acadian Parishes." In *The Cajuns: Essays on Their History and Culture*, ed. Glenn R. Conrad. Lafayette: Univ. of Southwestern Louisiana Center for Louisiana Studies.

Howard, Perry H. 1971. *Political Tendencies in Louisiana.* Baton Rouge: Louisiana State Univ. Press.

Kniffen, Fred B. 1968. *Louisiana: Its Land and People.* Baton Rouge: Louisiana State Univ. Press.

Lathrop, Barnes F. 1960. "The Lafourche District in 1861–1862: A Problem in Local Defense." *Louisiana History* 1: 99–129.

LeBlanc, Robert G. 1962. "Acadian Migrations." *Proceedings of the Minnesota Academy of Science* 30: 55–59.

Louder, Dean R., and Michael LeBlanc. 1979. "The Cajuns of East Texas." *Cahiers de Geographie du Quebec* 23: 317–30.

Martin, Paulette, trans. 1976. "The Kelly-Nugent Report of the Inhabitants and Livestock in the Attakapas, Natchitoches, Opelousas and Rapides Posts, 1770." *Attakapas Gazette* 6: 187–93.

Millet, Donald J. 1964. "The Economic Development of Southwest Louisiana, 1865–1900." Ph.D. diss., Louisiana State Univ.

Olmsted, Frederick L. 1953. *The Cotton Kingdom: A Traveller's Observation on Cotton and Slavery in the American Slave States,* ed. Arthur M. Schlesinger. New York: Knopf.

Oster, Harry. 1959. "Acculturation in Cajun Folk Music." *McNeese Review* 11: 12–24.

Oukada, Larbi. 1978. "The Territory and Population of French-Speaking Louisiana." *Revue de Louisiane* 7: 5–34.

Parenton, Vernon J. 1949. "Socio-Psychological Integration in a Rural French-Speaking Section of Louisiana." *Southwestern Social Science Quarterly* 30: 188–95.

Post, Lauren C. 1957. "The Old Cattle Industry of Southwest Louisiana." *McNeese Review* 9: 43–55.

Post, Lauren C. 1962. *Cajun Sketches from the Prairies of Southwest Louisiana.* Baton Rouge: Louisiana State Univ. Press.

Reed, Merl E. 1966. *New Orleans and the Railroads: The Struggle for Commercial Empire, 1830–1860.* Baton Rouge: Louisiana State Univ. Press.

Robison, R. Warren. 1975. "Louisiana Acadian Domestic Architecture." In *The Culture of Acadiana: Tradition and Change in South Louisiana,* ed. Steven L. Del Sesto and Jon L. Gibson. Lafayette: Univ. of Southwestern Louisiana Press.

Rushton, William F. 1979. *The Cajuns: From Acadia to Louisiana.* New York: Farrar, Straus & Giroux.

Sandoz, William J. 1925. "A Brief History of St. Landry Parish." *Louisiana Historical Quarterly* 8: 221–39.

Saucier, Corinne L. 1951. "A Historical Sketch of the Acadians." *Louisiana Historical Quarterly* 34: 63–89.

Scarpaci, Jean A. 1972. "Italian Immigrants in Louisiana's Sugar Parishes, 1880–1910." Ph.D. diss., Tulane Univ.

Shugg, Roger W. 1936. "Suffrage and Representation in Ante-Bellum Louisiana." *Louisiana Historical Quarterly* 19: 390–406.

Smith, Harley, and Hosea Phillips. 1939. "The Influence of English on Louisiana 'Cajun' French in Evangeline Parish." *American Speech* 14: 198–201.

Taylor, James W. 1950. "Louisiana Land Survey Systems." *Southwestern Social Science Quarterly* 31: 275–82.

Taylor, Joe Gray. 1976. *Louisiana: A Bicentennial History*. New York: Norton.

Tinker, Edward L. 1933. "Bibliography of the French Newspapers and Periodicals of Louisiana." *Proceedings*, American Antiquarian Society, New Series, 42: 247–368.

Tregle, Joseph G., Jr. 1954. "Louisiana in the Age of Jackson: A Study in Ego-Politics." Ph.D. diss., Univ. of Pennsylvania.

Tregle, Joseph G., Jr. 1972. "Political Reinforcement of Ethnic Dominance in Louisiana, 1812–1845." In *The Americanization of the Gulf Coast, 1803–1850*, Proceedings of the Gulf Coast History and Humanities Conference, III. Pensacola, Fl.

Trépanier, Cécyle. 1986. "The Catholic Church in French Louisiana: An Ethnic Institution?" *Journal of Cultural Geography* 7: 58–66.

U.S. Department of the Interior. 1864. *Agriculture of the United States in 1860*. Washington, D.C.: Government Printing Office.

U.S. Department of the Interior. 1902. *Agriculture*, Census Reports, vol. 6, pt. 2, *Crops and Irrigation*. Washington, D.C.: U.S. Census Office.

U.S. Population Schedules. 1820. Fourth Census of the United States. Microcopy No. M33.

U.S. Population Schedules. 1860. Eighth Census of the United States. Microcopy No. M653.

U.S. Population Schedules. 1900. Twelfth Census of the United States. Microcopy No. T623.

Winters, John D. 1963. *The Civil War in Louisiana*. Baton Rouge: Louisiana State Univ. Press.

Winzerling, Oscar W. 1955. *Acadian Odyssey*. Baton Rouge: Louisiana State Univ. Press.

CHAPTER 7: *La Tierra Tejana*

Adler, Jerry, and Tim Padgett. 1995. "Mexamerica: Selena Country." *Newsweek*, 23 October, 76–84.

Anders, Evan. 1982. *Boss Rule in South Texas: The Progressive Era*. Austin: Univ. of Texas Press.

Arreola, Daniel D. 1987. "The Mexican American Cultural Capital." *Geographical Review* 77, no. 1: 17–34.

Arreola, Daniel D. 1992. "Plaza Towns of South Texas." *Geographical Review* 81, no. 1: 56–73.

Arreola, Daniel D. 1993a. "Texas." In *Encyclopedia of American Social History*, ed. M. K. Clayton, E. J. Gorn, and P. W. Williams, 2: 1069–77. New York: Scribner's.

Arreola, Daniel D. 1993b. "The Texas-Mexican Homeland." *Journal of Cultural Geography* 13, no. 2: 61–74.

Arreola, Daniel D. 1993c. "Mexico Origins of South Texas Mexican Americans, 1930." *Journal of Historical Geography* 19, no. 1: 48–63.

Arreola, Daniel D. 1993d. "Plazas of San Diego, Texas: Signatures of Mexican-American Place Identity." *Places: A Quarterly Journal of Environmental Design* 8 (Spring): 80–87.

Arreola, Daniel D. 1993e. "Beyond the Nueces: The Early Geographical Identity of South Texas." *Río Grande: A Journal of Research and Issues* 2, no. 2: 46–60.

Arreola, Daniel D. 1995a. "Mexican Texas: A Distinctive Borderland." In *A Geographic Glimpse of Central Texas and the Borderlands: Images and Encounters*, ed. James F. Petersen and Julie A. Tuason, 3–9. Indiana, Pa.: National Council for Geographic Education.

Arreola, Daniel D. 1995b. "Urban Ethnic Landscape Identity." *Geographical Review* 85, no. 4: 518–34.

Arreola, Daniel D. 2000. "Mexican Americans." In *Ethnicity in Contemporary America: A Geographical Appraisal*, ed. Jesse O. McKee, 111–38. 2d ed. Lanham, Md.: Rowman & Littlefield.

Barbee, William Clayton. 1981. "A Historical and Architectural Investigation of San Ygnacio, Texas." M.A. thesis, Univ. of Texas, Austin.

Benavides, Adán, Jr. 1996. "Tejano." In *The New Handbook of Texas*, ed. Ron Tyler, 6: 238–39. 6 vols. Austin: Texas State Historical Association.

Blair, W. Frank. 1950. "The Biotic Provinces of Texas." *Texas Journal of Science* 2, no. 1: 93–117.

Boswell, Thomas D. 1979. "The Growth and Proportional Redistribution of the Mexican Stock Population in the United States: 1900–1970." *Mississippi Geographer* 6 (Spring): 57–76.

Brischetto, Robert R. 1988. "Electoral Empowerment: The Case for Tejanos." In *Latino Empowerment: Progress, Problems, and Prospects*, ed. Roberto E. Villareal, Norma G. Hernandez, and Howard D. Neighbor, 71–90. New York: Greenwood Press.

Byfield, Patsy Jeanne. 1966. *Falcón Dam and the Lost Towns of Zapata*. Austin: Univ. of Texas, Texas Memorial Museum.

Cardoso, Lawrence A. 1980. *Mexican Emigration to the United States 1897–1931: Socioeconomic Patterns*. Tucson: Univ. of Arizona Press.

Crouch, Dora P., Daniel J. Garr, and Alex I. Mundigo. 1982. *Spanish City Planning in North America*. Cambridge, Mass.: MIT Press.

Cruz, Gilbert R. 1988. *Let There Be Towns: Spanish Municipal Origins in the American Southwest, 1610–1810*. College Station: Texas A&M Univ. Press.

De León, Arnoldo. 1982. *The Tejano Community, 1836–1900*. Albuquerque: Univ. of New Mexico Press.

DeSipio, Louis. 1993. "The Dynamics of Mexican American Electoral Participation in South Texas." *South Texas Studies*, 89–121. Victoria, Tex.: Victoria College Press.

Fish, Jean Y. 1990. *Zapata County Roots Revisited*. Edinburg, Tex.: New Santander Press.

Fishman, Joshua. 1985. "The Ethnic Revival in the United States: Implications for the Mexican-American Community." In *Mexican-Americans in Comparative Perspective*, ed. Walker Conner, 311–54. Washington, D.C.: Urban Institute Press.

Foley, Douglas E., et al. 1977. *From Peones to Politicos: Ethnic Relations in a South Texas Town, 1900 to 1977*. Austin: Univ. of Texas, Center for Mexican American Studies.

Frantz, Joe B., and Mike Cox. 1988. *Lure of the Land: Texas County Maps and the History of Settlement*. College Station: Texas A&M Univ. Press.

Gamio, Manuel. 1930. *Mexican Immigration to the United States: A Study of Human Migration and Adjustment*. Chicago: Univ. of Chicago Press.

García, Ignacio M. 1989. *United We Win: The Rise and Fall of La Raza Unida Party*. Tucson: Univ. of Arizona Mexican American Studies Research Center.

García, Richard. 1991. *Rise of the Mexican American Middle Class: San Antonio, 1929–1941*. College Station: Texas A&M Univ. Press.

Garrett, Wilbur E., ed. 1988. "A Grand Fiesta Called Hispanics." In *Historical Atlas of the United States*, Centennial Edition, 72–73. Washington, D.C.: National Geographic Society.

George, Eugene. 1975. *Historic Architecture of Texas: The Falcon Reservoir*. Austin: Texas Historical Commission and Texas Historical Foundation.

Gonzalez, Jovita. 1930. "America Invades the Border Towns." *Southwest Review* 15: 469–477.

Graham, Joe S. 1985. "Folk Medicine and Intercultural Diversity among West Texas Mexican Americans." *Western Folklore* 44 (July): 168–93.

Graham, Joe S. 1991. "*Vaquero* Folk Arts and Crafts in South Texas." In *Hecho en Tejas: Texas-Mexican Folk Arts and Crafts*, ed. Joe S. Graham, 93–116. Denton: Univ. of North Texas Press.

Graham, Joe S. 1992. "The Built Environment in South Texas: The Hispanic Legacy." In *Hispanic Texas: A Historical Guide*, ed. Helen Simmons and Cathryn A. Hoyt, 58–75. Austin: Univ. of Texas Press.

Graham, Joe S. 1994. *El Rancho in South Texas: Continuity and Change from 1750*. Denton: John E. Conner Museum and Univ. of North Texas Press.

Haverluk, Terrence W. 1993. "Mex-America: The Maintenance and Expansion of an American Cultural Region." Ph.D. diss., Univ. of Minnesota.

Hufford, Larry. 1988. "The Velásquez Legacy." *Texas Observer*, 29 July, 6–8.

Institute of Texan Cultures. 1971. *Los Mexicano Texanos*. San Antonio: Univ. of Texas, Institute of Texan Cultures.

Jackson, Jack. 1986. *Los Mesteños: Spanish Ranching in Texas, 1721–1821*. College Station: Texas A&M Univ. Press.

Jordan, Terry G. 1966. *German Seed in Texas Soil: Immigrant Farmers in Nineteenth-Century Texas*. Austin: Univ. of Texas Press.

Jordan, Terry G. 1978. "Perceptual Regions in Texas." *Geographical Review* 68, no. 3: 293–307.

Jordan, Terry G. 1986. "A Century and a Half of Ethnic Change in Texas, 1836–1986." *Southwestern Historical Quarterly* 89 (April): 385–422.

Jordan, Terry G. 1993. *North American Cattle-Ranching Frontiers: Origins, Diffusion, and Differentiation.* Albuquerque: Univ. of New Mexico Press.

Jordan, Terry G., with John L. Bean Jr. and William M. Holmes. 1984. *Texas: A Geography.* Boulder: Westview Press.

Juárez, José Roberto. 1973. "La Iglesia Católica y el Chicano en Sud Texas 1836–1911." *Aztlán* 4, no. 2: 217–55.

Kelly, Pat. 1986. *River of Lost Dreams: Navigation on the Rio Grande.* Lincoln: Univ. of Nebraska Press.

Kibbe, Pauline R. 1946. *Latin Americans in Texas.* Albuquerque: Univ. of New Mexico Press.

La Prensa, 1923. Anniversary Edition [San Antonio] Feb. 13.

Larralde, Carlos Montalvo. 1978. "Chicano Jews in South Texas." Ph.D. diss., Univ. of California, Los Angeles.

Lewis, Peirce F. 1979. "Axioms for Reading the Landscape: Some Guides to the American Scene." In *The Interpretation of Ordinary Landscapes: Geographical Essays,* ed. D. W. Meinig, 11–32. New York: Oxford Univ. Press.

Liebman, Seymour B. 1970. *The Jews in New Spain: Faith, Flame, and the Inquisition.* Coral Gables, Fla.: Univ. of Miami Press.

Lott, Virgil N., and Mercurio Martínez. 1953. *The Kingdom of Zapata.* Austin: Eakin Press.

Madsen, William. 1964. *The Mexican-Americans of South Texas* [Hidalgo County]. New York: Holt, Rinehart, & Winston.

Maril, Robert Lee. 1992. *Living on the Edge of America: At Home on the Texas-Mexico Border.* College Station: Texas A&M Univ. Press.

Markides, Kyriakos, and Thomas Cole. 1984. "Change and Continuity in Mexican American Religious Behavior: A Three-Generation Study." *Social Science Quarterly* 65, no. 2: 618–25.

Márquez, Benjamin. 1993. *LULAC: The Evolution of a Mexican American Political Organization.* Austin: Univ. of Texas Press.

Martínez, M. J. 1873. "Mapa del Rio Grande." In *Reports of the Committee of Investigation Sent in 1873 by the Mexican Government to the Frontier of Texas.* [Translated from the official edition.] New York: Baker & Godwin, 1875.

Matovina, Timothy M. 1995. *Tejano Religion and Ethnicity: San Antonio, 1821–1860.* Austin: Univ. of Texas Press.

McCleskey, Clifton, and Bruce Merrill. 1973. "Mexican-American Political Behavior in Texas." *Social Science Quarterly* 53, no. 1: 785–98.

McVey, Lori Brown. 1988. *Guerrero Viejo: A Photographic Essay.* Laredo, Tex.: Nuevo Santander Museum.

Meier, Matt S., and Feliciano Rivera. 1981. *Dictionary of Mexican-American History.* Westport, Conn.: Greenwood Press.

Meinig, Donald W. 1969. *Imperial Texas: An Interpretive Essay in Cultural Geography.* Austin: Univ. of Texas Press.

Miller, Michael V. 1975. "Chicano Community Control in South Texas: Problems and Prospects." *Journal of Ethnic Studies* 3 (Fall): 70–89.

Montano, Mario. 1992. "The History of Mexican Folk Foodways of South Texas: Street Vendors, Offal Foods, and *Barbacoa de Cabeza.*" Ph.D. diss., Univ. of Pennsylvania.

Montejano, David. 1987. *Anglos and Mexicans in the Making of Texas, 1836–1986.* Austin: Univ. of Texas Press.

National Association of Latino Elected and Appointed Officials. 1984–94. *National Roster of Hispanic Elected Officials.* Washington, D.C.: NALEO Education Fund.

Newton, Ada L. 1964. "The History of Architecture along the Rio Grande as Reflected in the Buildings around Rio Grande City, 1749–1920." M.A. thesis, Texas A&I, Kingsville.

Nostrand, Richard L. 1992a. *The Hispano Homeland.* Norman: Univ. of Oklahoma Press.

Nostrand, Richard L. 1992b. "Greater New Mexico's Hispano Island." *Focus* 42, no. 4: 13–19.

O'Shea, Elena Zamora. 1935. *El Mesquite: A Story of the Early Spanish Settlements between the Nueces and the Rio Grande as Told by "La Posta del Palo Alto."* Dallas: Mathis Publishing.

Paredes, Américo. 1993. "The Problem of Identity in a Changing Culture: Popular Expressions of Culture Conflict along the Lower Rio Grande Border." In *Folklore and Culture on the Texas-Mexican Border,* ed. Richard Bauman, 19–48. Austin: Univ. of Texas, Center for Mexican American Studies.

Peña, Manuel. 1985. *The Texas-Mexican Conjunto: History of a Working-Class Music.* Austin: Univ. of Texas Press.

Pierce, Frank Cushman. 1917. *A Brief History of the Lower Rio Grande.* Menasha, Wis.: George Banta Publishing.

Poyo, Gerald E., and Gilberto M. Hinojosa, eds. 1991. *Tejano Origins in Eighteenth-Century San Antonio.* Austin: Univ. of Texas Press.

Remy, Martha C. M. 1970. "Protestant Churches and Mexican-Americans in South Texas." Ph.D. diss., Univ. of Texas, Austin.

Reps, John W. 1979. *Cities of the American West: A History of Frontier Urban Planning.* Princeton: Princeton Univ. Press.

Roberts, Charles. 1995. "*Imperial Texas:* Texan Myths and Tejano Exclusion." *Historical Geography* 24, nos. 1 and 2: 45–56.

Robinson, Willard B. 1979. "Colonial Ranch Architecture in the Spanish-Mexican Tradition." *Southwestern Historical Quarterly* 83 (October): 123–50.

Romano V., Octavio Ignacio. 1965. "Charismatic Medicine, Folk-Healing, and Folk-Sainthood." *American Anthropologist* 67: 1151–73.

Sánchez, Mario L. 1991. *A Shared Experience: The History, Architecture and Historic Des-*

ignations of the Lower Rio Grande Heritage Corridor. Austin: Los Caminos del Rio and the Texas Historical Commission.

Sánchez, Ramón. 1898. *Un viaje de maravatio a San Antonio de Bejar (Texas).* Zamora, Mex.: Tip. Moderna.

Sandos, James A. 1992. *Rebellion in the Borderlands: Anarchism and the Plan of San Diego, 1904–1923.* Norman: Univ. of Oklahoma Press.

Shelly, Fred M., J. Clark Archer, and G. Thomas Murauskas. 1986. "The Geography of Recent Presidential Elections in the Southwest." *Arkansas Journal of Geography* 2: 1–9.

Shelton, Edgar G., Jr. 1974. *Political Conditions among Texas Mexicans along the Rio Grande.* San Francisco: R&E Research Associates.

Shockley, John Staples. 1974. *Chicano Revolt in a Texas Town.* Notre Dame: Univ. of Notre Dame Press.

Shortridge, James R. 1976. "Patterns of Religion in the United States." *Geographical Review* 66, no. 4: 420–34.

Sologaistoa, J. C., ed. 1924. *Guia general y directorio mexicano de San Antonio, Texas.* San Antonio: San Antonio Paper Co.

Talbert, Robert H. 1955. *Spanish-Name People in the Southwest and West: Socio-economic Characteristics of White Persons of Spanish Surname in Texas, Arizona, California, Colorado, and New Mexico.* Fort Worth: Leo Potishman Foundation and Texas Christian Univ.

Taylor, Paul Schuster. 1934. *An American-Mexican Frontier: Nueces County, Texas.* Chapel Hill: Univ. of North Carolina Press.

U.S. Bureau of the Census. 1990a. *Census of Population and Housing, Texas.* Washington, D.C.: Government Printing Office, 1991.

U.S. Bureau of the Census. 1990b. *Census of Population and Housing,* Summary Tape File 3C. Washington, D.C.: Government Printing Office.

U.S. Bureau of the Census, 2000. *2000 Census of Population and Housing.* www.census.gov.

West, Richard. 1981. *Richard West's Texas.* Austin: Texas Monthly Press.

Wilkinson, J. B. 1975. *Laredo and the Rio Grande Frontier.* Austin: Jenkins Publishing.

CHAPTER 8: *The Anglo-Texan Homeland*

Carlson, Alvar W. 1990. *The Spanish-American Homeland: Four Centuries in New Mexico's Río Arriba.* Baltimore: Johns Hopkins Univ. Press.

Fehrenbach, T. R. 1974. *Comanches: The Destruction of a People.* New York: Knopf.

Foster, L. L., ed. 1889. *First Annual Report of the Agricultural Bureau of the Department of Agriculture, Insurance, Statistics, and History, 1887–88.* Austin: State Printing Office.

Garcia, James E. 1991. "Minorities in Texas' Schools Are the Majority." *Austin American-Statesman* 121, no. 44: A1, A6.

Jordan, Terry G. 1986. "A Century and a Half of Ethnic Change in Texas, 1836–1986." *Southwestern Historical Quarterly* 89: 385–422.

Jordan, Terry G. 1993a. *North American Cattle-Ranching Frontiers.* Albuquerque: Univ. of New Mexico Press.

Jordan, Terry G. 1993b. "The Anglo-Texan Homeland." *Journal of Cultural Geography* 13, no. 2: 75–86.

Kilgore, Dan E. 1978. *How Did Davy Die?* College Station: Texas A&M Univ. Press.

Kingston, Mike, ed. 1991. *1992–1993 Texas Almanac and State Industrial Guide.* Dallas: Dallas Morning News.

Meinig, Donald W. 1969. *Imperial Texas: An Interpretive Essay in Cultural Geography.* Austin: Univ. of Texas Press.

Meinig, Donald W. 1971. *Southwest: Three Peoples in Geographical Change, 1600–1970.* New York: Oxford Univ. Press.

Nostrand, Richard L. 1980. "The Hispano Homeland in 1900." *Annals of the Association of American Geographers* 70: 382–96.

Nostrand, Richard L. 1992. *The Hispano Homeland.* Norman: Univ. of Oklahoma Press.

Peña, Manuel H. 1985. *The Texas-Mexican Conjunto: History of a Working-Class Music.* Austin: Univ. of Texas Press.

Reed, John S. 1976. "The Heart of Dixie: An Essay in Folk Geography." *Social Forces* 54: 923–39.

Shockley, John S. 1974. *Chicano Revolt in a Texas Town.* Notre Dame: Univ. of Notre Dame Press.

Webb, Walter P., and H. Bailey Carroll, eds. 1952. *The Handbook of Texas.* 2 vols. Austin: Texas State Historical Association.

Zelinsky, Wilbur. 1980. "North America's Vernacular Regions." *Annals of the Association of American Geographers* 70: 1–16.

CHAPTER 9: *The Kiowa Homeland in Oklahoma*

Boyd, Maurice. 1981. *Kiowa Voices,* vol. 1: *Ceremonial Dance, Ritual and Song.* Fort Worth: Texas Christian Univ. Press.

Boyd, Maurice. 1983. *Kiowa Voices,* vol. 2: *Myths, Legends and Folktales.* Fort Worth: Texas Christian Univ. Press.

Bureau of Indian Affairs. ca. 1901. Allotment Records for Apache, Comanche, Kiowa, Caddo, Wichita, and Affiliated Bands. Microfilm roll KA 89, Oklahoma Historical Society, Oklahoma City.

Corwin, Hugh D. 1968. "Protestant Missionary Work among the Comanches and Kiowas." *Chronicles of Oklahoma* 46, no. 1: 41–57.

Ellis, Clyde. 1990. "'Truly Dancing Their Own Way': Modern Revival and Diffusion of the Gourd Dance." *American Indian Quarterly* 14, no. 1: 19–33.

Griffin, Abial W. 1901. Allotment Map of the Kiowa, Comanche, and Apache Indian Reservation in Oklahoma. Butte, Okla.: Griffin, Davis & Thrush. From Oklahoma Historical Society, Oklahoma City.

Hagan, William T. 1990. *United States–Comanche Relations: The Reservation Years.* 2d ed. Norman: Univ. of Oklahoma Press.

Mayhall, Mildred P. 1971. *The Kiowas.* 2d ed. Norman: Univ. of Oklahoma Press.

Momaday, N. Scott. 1969. *The Way to Rainy Mountain.* Albuquerque: Univ. of New Mexico Press.

Momaday, N. Scott. 1989. *The Ancient Child.* New York: Doubleday.

Mooney, James. 1979. *Calendar History of the Kiowa Indians.* Reprint of 1898 edition, Bureau of American Ethnology Annual Report 17, pt. 2. Washington, D.C.: Smithsonian Institution Press.

Nye, Wilbur S. 1942. *Carbine and Lance: The Story of Old Fort Sill.* 2d ed. Norman: Univ. of Oklahoma Press.

RMKIBC (Rainy Mountain Kiowa Indian Baptist Church). 1993. *Celebrating 100 Years— 1893–1993.* Mt. View, Okla.: Rainy Mountain Kiowa Indian Baptist Church.

Vernon, Walter N. 1980–81. "Methodist Beginnings among Southwest Oklahoma Indians." *Chronicles of Oklahoma* 58, no. 4: 392–411.

CHAPTER 10: *The Highland-Hispano Homeland*

Barreiro, Antonio. 1928. "Barreiro's Ojeada Sobre Nuevo-Mexico" (1832), trans. and ed. Lansing B. Bloom. *New Mexico Historical Review* 3, no. 1: 73–96.

Bloom, Lansing Bartlett. 1913. "New Mexico under Mexican Administration, 1821–1846." *Old Santa Fe* 1, no. 1: 3–49.

Carlson, Alvar W. 1975. "Long Lots in the Rio Arriba." *Annals of the Association of American Geographers* 65, no. 1: 48–57.

Chávez, Fray Angélico. 1949. "Saints' Names in New Mexico Geography." *El Palacio* 56, no. 11: 323–35.

Chávez, Fray Angélico. 1950. "New Mexico Religious Place-Names Other Than Those of Saints." *El Palacio* 57, no. 1: 23–26.

Chávez, Fray Angélico. 1953. "The 'Kingdom of New Mexico.'" *New Mexico Magazine* 31: no. 8: 17, 58–59; no. 9: 17, 42–43.

De Borhegyi, Stephen F., and E. Boyd. 1956. *El Santuario de Chimayo.* Reprint, 1987. Santa Fe: Ancient City Press.

Deutsch, Sarah. 1987. *No Separate Refuge: Culture, Class, and Gender on an Anglo-Hispanic Frontier in the American Southwest, 1880–1940.* New York: Oxford Univ. Press.

Ellis, Florence Hawley. 1987. "The Long Lost 'City' of San Gabriel del Yunge, Second Oldest European Settlement in the United States." In *When Cultures Meet: Remembering San Gabriel del Yunge Oweenge,* ed. Herman Agoyo and Lynnwood Brown, 10–38. Santa Fe: Sunstone Press.

Foster, George M. 1960. *Culture and Conquest: America's Spanish Heritage.* Chicago: Quadrangle Books, Viking Fund Publications in Anthropology no. 27.

Gregg, Josiah. 1844. *Commerce of the Prairies*, ed. Max L. Moorhead. Reprint, 1954. Norman: Univ. of Oklahoma Press.

Gritzner, Charles F. 1971. "Log Housing in New Mexico." *Pioneer America* 3, no. 2: 54–62.

Gritzner, Charles. 1974a. "Construction Materials in a Folk Housing Tradition: Considerations Governing Their Selection in New Mexico." *Pioneer America* 6, no. 1: 25–39.

Gritzner, Charles F. 1974b. "Hispano Gristmills in New Mexico." *Annals of the Association of American Geographers* 64, no. 4: 514–24.

Jordan, Terry G. 1974. "Antecedents of the Long Lot in Texas." *Annals of the Association of American Geographers* 64, no. 1: 70–86.

Kluckhohn, Florence Rockwood, and Fred L. Strodtbeck. 1961. *Variations in Value Orientations*. Evanston, Ill.: Row, Peterson.

Leonard, Olen, and C. P. Loomis. 1941. *Culture of a Contemporary Rural Community: El Cerrito, New Mexico*. Rural Life Studies 1. Washington, D.C.: U.S. Department of Agriculture.

Mead, Margaret, ed. 1955. *Cultural Patterns and Technical Change*. New York: New American Library.

Meinig, D. W. 1971. *Southwest: Three Peoples in Geographical Change, 1600–1970*. New York: Oxford Univ. Press.

Pino, Pedro Bautista. 1942. *The* Exposición *of Don Pedro Bautista Pino 1812*, trans. and ed. H. Bailey Carroll and J. Villasana Haggard, vol. 11. Reprint. Albuquerque: Quivera Society.

Pitt, Leonard. 1966. *The Decline of the Californios: A Social History of the Spanish-Speaking Californians, 1846–1890*. Berkeley: Univ. of California Press.

Ruxton, George Frederick. 1848. *Wild Life in the Rocky Mountains*. Reprint, 1916. New York: Macmillan.

Sporleder, Louis B. 1933. "La Plaza de los Leones." *Colorado Magazine* 10, no. 1: 28–38.

Sunseri, Alvin R. 1973. "Agricultural Techniques in New Mexico at the Time of the Anglo-American Conquest." *Agricultural History* 47, no. 4: 329–37.

Tuan, Yi-Fu, and Cyril E. Everard. 1964. "New Mexico's Climate: The Appreciation of a Resource." *Natural Resources Journal* 4, no. 2: 268–308.

Weber, David J. 1985. Comments as a panelist in The Spanish Americans: Independent Cultural Groupings or Mexican American Subculture? Session, Western Social Science Association, Fort Worth, Texas, 27 April 1985.

Winberry, John J. 1974. "The Log House in Mexico." *Annals of the Association of American Geographers* 64, no. 1: 54–69.

Winberry, John J. 1975. "*Tejamanil:* The Origin of the Shake Roof in Mexico." *Proceedings of the Association of American Geographers* 7: 288–93.

Zunser, Helen. 1935. "A New Mexican Village." *Journal of American Folklore* 48 (April–June): 125–78.

CHAPTER 11: *The Navajo Homeland*

Astrov, Margot. 1950. "The Concept of Motion as a Psychological Leitmotif of Navaho Life and Literature." *Journal of American Folklore* 63, no. 247: 45–56.

Bailey, Lynn R. 1964. *The Long Walk: A History of the Navajo Wars, 1846–1868*. Pasadena: Westernlore Publications.

Beck, Peggy V., and Anna L. Walters. 1977. *The Sacred: Ways of Knowledge, Sources of Life*. Tsaile, Ariz.: Navajo Community College Press.

Bingham, Sam, and Janet Bingham. 1979. *Navajo Farming*. Rock Point, Ariz.: Rock Point Community School.

Bingham, Sam, and Janet Bingham. 1987. *Navajo Chapters*. Rev. ed. Tsaile, Ariz.: Navajo Community College Press.

Blake, Kevin. 1999. "Sacred and Secular Landscape Symbolism at Mount Taylor, New Mexico." *Journal of the Southwest* 41, no. 4: 487–509.

Brugge, David M. 1968. "Pueblo Influence on Navajo Architecture." *El Palacio* 75, no. 3: 14–20.

Brugge, David M. 1983. "Navajo History and Prehistory to 1850." In *Handbook of North American Indians*, vol. 10: *Southwest*, ed. Alfonso Ortiz, 489–501. Washington, D.C.: Smithsonian Institution.

Brugge, David M. 1994. *The Navajo-Hopi Land Dispute: An American Tragedy*. Albuquerque: Univ. of New Mexico Press.

Brugge, David M., and J. Lee Correll. 1971. *The Story of the Navajo Treaties*. Navajo Historical Publications, Documentary Series 1.

Clemmer, Richard O. 1995. *Roads in the Sky: The Hopi Indians in a Century of Change*. Boulder: Westview Press.

Comeaux, Malcolm. 1992. "Cajuns in Louisiana." In *To Build in a New Land: Ethnic Landscapes in North America*, ed. Allen G. Noble, 177–92. Baltimore: Johns Hopkins Univ. Press.

Correll, J. Lee, and Alfred Dehiya. 1972. *Anatomy of the Navajo Indian Reservation: How It Grew*. Window Rock, Ariz.: Navajo Times Publishing Co. Rev. ed., 1978; reprinted 1995 by the Navajo Nation, Window Rock, Arizona; text taken from *Navajo Tribal Code*, Appendix.

Emerson, Gloria J. 1983. "Navajo Education." In *Handbook of North American Indians*, vol. 10: *Southwest*, ed. Alfonso Ortiz, 659–71. Washington, D.C.: Smithsonian Institution.

Farella, John R. 1984. *The Main Stalk: A Synthesis of Navajo Philosophy*. Tucson: Univ. of Arizona Press.

Frisbie, Charlotte J. 1992. "Temporal Change in Navajo Religion." *Journal of the Southwest* 34, no. 4: 457–514.

Gold, Peter. 1994. *Navajo and Tibetan Sacred Wisdom: The Circle of the Spirit*. Rochester, Vt.: Inner Traditions.

Goodman, James M. 1982. *The Navajo Atlas: Environments, Resources, People, and History of the Diné Bikeyah.* Norman: Univ. of Oklahoma Press.

Griffin-Pierce, Trudy. 1992. *Earth Is My Mother, Sky Is My Father: Space, Time, and Astronomy in Navajo Sandpainting.* Albuquerque: Univ. of New Mexico Press.

Grinde, Donald, and Bruce E. Johansen. 1995. *Ecocide of Native America: Environmental Destruction of Native Lands.* Santa Fe: Clear Light Publishers.

Haile, Berard. 1943. *Soul Concepts of the Navajo.* Vatican City: Tipografia del Vaticano [orig. pub. in *Annali Lateranensi* 7: 59–94].

Harvey, Sioux. 1996. "Two Models to Sovereignty: A Comparative History of the Mashantucket Pequot Tribal Nation and the Navajo Nation." *American Indian Culture and Research Journal* 20, no. 1: 147–94.

Hester, James J. 1962. *Early Navajo Migrations and Acculturations in the Southwest.* Museum of New Mexico Papers in Anthropology 6.

Iverson, Peter. 1981. *The Navajo Nation.* Westport, Conn.: Greenwood Press.

Iverson, Peter. 1983. "The Emerging Navajo Nation." In *Handbook of North American Indians*, vol. 10: *Southwest*, ed. Alfonso Ortiz, 636–40. Washington, D.C.: Smithsonian Institution.

Jett, Stephen C. 1970. "An Analysis of Navajo Place-Names." *Names* 18, no. 3: 175–84.

Jett, Stephen C. 1978a. "Navajo Seasonal Migration Patterns." *Kiva* 44, no. 1: 65–75.

Jett, Stephen C. 1978b. "Origin of Navajo Settlement Patterns." *Annals of the Association of American Geographers* 68, no. 3: 351–62.

Jett, Stephen C. 1980. "The Navajo Homestead: Situation and Site." *Yearbook of the Association of Pacific Coast Geographers* 41: 101–18.

Jett, Stephen C. 1990. "Culture and Tourism in the Navajo Country." *Journal of Cultural Geography* 11, no. 1: 85–107.

Jett, Stephen C. 1992a. "An Introduction to Navajo Sacred Places." *Journal of Cultural Geography* 13, no. 1: 29–39.

Jett, Stephen C. 1992b. "The Navajo in the American Southwest." In *To Build in a New Land: Ethnic Landscapes in North America*, ed. Allen G. Noble, 331–44. Baltimore: Johns Hopkins Univ. Press.

Jett, Stephen C. 1995a. "Cultural Fusion in Native-American Folk Architecture: The Navajo Hogan." In *American Indians: A Cultural Geography*, ed. Thomas G. Ross, Tyrel G. Moore, and Laura R. King, pp. 227–40. 2d ed. Southern Pines, N.C.: Karo Hollow Press. Originally published 1987 in *A Cultural Geography of North American Indians*, ed. Thomas G. Ross and Tyrel G. Moore, pp. 243–56. Boulder: Westview Press.

Jett, Stephen C. 1995b. "Navajo Sacred Places: The Management and Interpretation of Mythic History." *Public Historian* 17, no. 2: 39–47.

Jett, Stephen C. 1997. "Place-Naming, Environment, and Perception among the Canyon de Chelly Navajo." *Professional Geographer* 49, no. 4: 481–93.

Jett, Stephen C. 1998a. "Scenic Resources and Tourism Development in the Navajo

Country." In *Tourism and Gaming on American Indian Lands*, ed. Alan Lew and George A. Van Otten, 93–110. Elmsford, N.Y.: Cognizant Communication.

Jett, Stephen C. 1998b. "Territory and Hogan: Local Homelands of the Navajo." In *Diné Bíkeyah: Papers in Honor of David M. Brugge*, ed. Meliha S. Duran and David T. Kirkpatrick, 117–28. Archaeological Society of New Mexico Papers, 24. Albuquerque.

Jett, Stephen C. 2001. *Navajo Placenames and Trails of the Canyon de Chelly System, Arizona*. New York: Peter Lang.

Jett, Stephen C., and Virginia E. Spencer. 1981. *Navajo Architecture: Forms, History, Distributions*. Tucson: Univ. of Arizona Press.

Johnson, Trebbe. 1987. "Between Sacred Mountains: The Hopi and Navajo Concept of Place." *Amicus Journal* 9, no. 3: 18–23.

Kammer, Jerry. 1980. *The Second Long Walk: The Navajo-Hopi Land Dispute*. Albuquerque: Univ. of New Mexico Press.

Kelley, Klara B. 1986. *Navajo Land Use: An Ethnoarchaeological Study*. Orlando: Academic Press.

Kelley, Klara Bonsack, and Harris Francis. 1994. *Navajo Sacred Places*. Bloomington: Indiana Univ. Press.

Kelley, Klara B., and Peter M. Whiteley. 1989. *Navajoland: Family and Settlement and Land Use*. Tsaile, Ariz.: Navajo Community College Press.

LaDuke, Winona. 1996. "Like Tributaries to a River." *Sierra* 81, no. 6: 38–45.

Lindig, Wolfgang, and Helga Teiwes. 1991. *Navajo: Tradition and Change in the Southwest*. New York: Facts on File.

Link, Martin A., ed. 1968. *Treaty between the United States of America and the Navajo Tribe of Indians*. Flagstaff: KC Publications.

Locke, Raymond Friday. 1979. *The Book of the Navajo*. Los Angeles: Mankind Publishing.

Lord, Nancy. 1996. "Native Tongues." *Sierra* 81, no. 6: 46–49, 68–69.

Martin, Rena. 1995. "Chapter Officials Tour Dinétah." *Navajo Preservation News*, Summer, 1, 2, 7–8.

McNitt, Frank. 1972. *Navajo Wars: Military Campaigns, Slave Raids, and Reprisals*. Albuquerque: Univ. of New Mexico Press.

Moore, William Haas. 1994. *Chiefs, Agents and Soldiers: Conflict on the Navajo Frontier, 1868–1882*. Albuquerque: Univ. of New Mexico Press.

Nostrand, Richard L. 1992. *The Hispano Homeland*. Norman: Univ. of Oklahoma Press.

Paisano, Edna L. 1993. *We the First Americans*. Washington, D.C.: U.S. Bureau of the Census.

Pavlik, Steve. 1990. "Climbing Manuelito's Ladder: Navajo Indian Education since 1868." *Diné Be'iina': A Journal of Navajo Life* 2, no. 1: 87–103.

Perry, Richard J. 1991. *Western Apache Heritage: People of the Mountain Corridor*. Austin: Univ. of Texas Press.

Pinxten, Rik, and Ingrid Van Dooren. 1992. "Navajo Earth and Sky and the Celestial Life Force." In *Earth and Sky: Visions of the Cosmos in Native American Folklore*, ed.

Ray A. Williams and Claire R. Farrer, 101–9. Albuquerque: Univ. of New Mexico Press.

Pollack, Floyd A. 1984. *A Navajo Confrontation and Crisis.* Tsaile, Ariz.: Navajo Community College Press.

Redhouse, John. 1984. *The Leasing of Dinetah.* Albuquerque: Redhouse/Wright Productions.

Reichard, Gladys A. 1963. *Navaho Religion: A Study of Symbolism.* Bollingen Series 18. New York: Pantheon Books. [Orig. pub. 1950]

Reid, Betty. 1992. "The Two Worlds of the Navajo." In Joel Grimes, *Navajo: Portrait of a Nation*, pp. 19–20, 22, 59–60, 93–94, 98, 100, 105, 121–22, 126, 128, 153–54, 156. Englewood, Colo.: Westcliffe Publishers.

Roessel, Monty. 1990. "The Dinetah-Navajo Sacred Homeland." In *Indians of New Mexico*, ed. Richard C. Sandoval and Ree Shenk, pp. 122–25. Santa Fe: New Mexico Magazine.

Roessel, Robert A., Jr. 1977. *Navajo Education in Action: The Rough Rock Demonstration School.* Chinle, Ariz.: Navajo Curriculum Center, Rough Rock Demonstration School.

Schwarz, Maureen Trudelle. 1997. "Unravelling the Anchoring Cord." *American Anthropologist* 99, no. 1: 43–55.

Shepardson, Mary. 1963. *Navajo Ways in Government: A Study in Political Process.* American Anthropological Association, Memoir 96. Menasha, Wis.

Shinkle, James D. 1965. *Fort Sumner and the Bosque Redondo Indian Reservation.* Roswell, N.M.: Hall-Poorbaugh Press.

Shumway, J. Matthew, and Richard H. Jackson. 1995. "Native American Population Patterns." *Geographical Review* 85, no. 2: 185–201.

Sleight, Frederick W. 1950. "The Navajo Sacred Mountain of the East—A Controversy." *El Palacio* 58, no. 12: 379–97.

Sutton, Imre. 1991. Preface to "Indian Country: Geography and Law." *American Indian Culture and Research Journal* 15, no. 2: 3–35.

Sutton, Imre. 1994. "Indian Land, White Man's Law: Southern California Revisited." *American Indian Culture and Research Journal* 28, no. 3: 265–70.

Thompson, Hildegard. 1975. *The Navajos' Long Walk for Education: A History of Navajo Education.* Tsaile, Ariz.: Navajo Community College Press.

Trafzer, Clifford E. 1982. *The Kit Carson Campaign: The Last Great Navajo War.* Norman: Univ. of Oklahoma Press.

Underhill, Ruth. 1956. *The Navajos.* Norman: Univ. of Oklahoma Press.

Van Valkenburgh, Richard F. 1941. *Diné Bikéyah.* Window Rock, Ariz.: Navajo Service.

Witherspoon, Gary. 1975. *Navajo Kinship and Marriage.* Chicago: Univ. of Chicago Press.

Witherspoon, Gary, and Glen Peterson. 1995. *Dynamic Symmetry and Holistic Asymmetry in Navajo and Western Art and Cosmology.* New York: Peter Lang Publishing.

Young, Robert W. 1961. *The Navajo Yearbook 8.* Window Rock, Ariz.: Navajo Agency.

Young, Robert W. 1978. *A Political History of the Navajo Tribe*. Tsaile, Ariz.: Navajo Community College Press.

Young, Robert W., and William Morgan. 1954. *Navajo Historical Selections*. Navajo Historical Series 3. Phoenix: Bureau of Indian Affairs.

CHAPTER 12: *Mormondom's Deseret Homeland*

Alexander, Thomas G. 1995a. *Utah, The Right Place: The Official Centennial History*. Layton, Utah: Gibbs Smith.

Alexander, Thomas G. 1995b. "The Emergence of a Republican Majority in Utah, 1970–1992." In *Politics in the Postwar American West*, ed. Richard Lowitt, 260–76. Norman: Univ. of Oklahoma Press.

Allen, James, and Glen Leonard. 1992. *The Story of the Latter-day Saints*. 2d ed. Salt Lake City: Deseret Book.

Arrington, Leonard J. 1958. *Great Basin Kingdom: An Economic History of the Latter-day Saints, 1830–1900*. Cambridge, Mass.: Harvard Univ. Press.

Barney, Ronald O. 1978. "The Life and Times of Lewis Barney." Master's thesis, Utah State Univ.

Barth, Gunther. 1975. *Instant Cities: Urbanization and the Rise of San Francisco and Denver*. New York: Oxford Univ. Press.

Bennett, Richard E. 1997. *We'll Find the Place: The Mormon Exodus, 1846–1848*. Salt Lake City: Deseret Book.

Bennion, John. 1994. "Doubt and the Desert." *Association for Mormon Letters Annual* 2: 263–69.

Bennion, Lowell C. 1991. "A Geographer's Discovery of 'Great Basin Kingdom.'" In *Great Basin Kingdom Revisited: Contemporary Perspectives*, ed. Thomas G. Alexander, 109–32. Logan: Utah State Univ. Press.

Bennion, Lowell C., and Lawrence A. Young. 1996. "The Uncertain Dynamics of LDS Expansion, 1950–2020." *Dialogue* 29, no. 1: 8–32.

Brown, S. Kent, et al., eds. 1994. *Historical Atlas of Mormonism*. New York: Simon & Schuster.

Bushman, Richard. 1997. *Making Space for the Mormons*. Logan: Utah State Univ. Press.

Conkin, Paul K. 1997. *American Originals: Homemade Varieties of Christianity*. Chapel Hill: Univ. of North Carolina Press.

De Pillis, Mario S. 1996. "The Emergence of Mormon Power since 1945." *Journal of Mormon History* 22, no. 1: 1–32.

Esplin, Ronald K. 1982. "'A Place Prepared': Joseph, Brigham and the Quest for Promised Refuge in the West." *Journal of Mormon History* 9: 85–111.

Geary, Edward A. 1996a. "Redeeming the Waste Places of Zion: The History of a Metaphor." P. A. Christensen Memorial Lecture, Provo, Utah, Brigham Young Univ.

Geary, Edward A. 1996b. *A History of Emery County*. Salt Lake City: Utah State Historical Society.

Hoskins, Shannon R., ed. 1996. *Faces of Utah: A Portrait.* Layton, Utah: Gibbs Smith.

Jackson, Richard H. 1981. "Utah's Harsh Lands, Hearth of Greatness." *Utah Historical Quarterly* 49, no. 1: 4–25.

Kay, Jeanne. 1995. "Mormons and Mountains." In *The Mountainous West: Explorations in Historical Geography,* ed. William Wyckoff and Lary M. Dilsaver, 368–95. Lincoln: Univ. of Nebraska Press.

Lyon, Thomas, and Terry Tempest Williams, eds. 1995. *Great and Peculiar Beauty: A Utah Reader.* Layton, Utah: Gibbs Smith.

Madsen, Brigham D., ed. 1989. *Exploring the Great Salt Lake: The Stansbury Expedition of 1849–50.* Salt Lake City: Univ. of Utah Press.

Mauss, Armand L. 1994. *The Angel and the Beehive.* Urbana: Univ. of Illinois Press.

May, Dean L. 1977. "The Making of Saints: The Mormon Town as a Setting for the Study of Cultural Change." *Utah Historical Quarterly* 45, no. 1: 75–92.

Meinig, D. W. 1965. "The Mormon Culture Region: Strategies and Patterns in the Geography of the American West, 1847–1964." *Annals of the Association of American Geographers* 55, no. 2: 191–220.

Meinig, D. W. 1997. "A Response." *Historical Geography* 25: 1–9.

Meinig, D. W. 1998. *The Shaping of America,* vol. 3. New Haven: Yale Univ. Press.

Morgan, Dale L. 1949. "Salt Lake City." In *Rocky Mountain Cities,* ed. Ray B. West, Jr., 179–207. New York: Norton.

Morgan, Dale L. 1953. *Jedediah Smith and the Opening of the West.* Indianapolis: Bobbs-Merrill.

Mulder, William. 1957. "The Mormons in American History." Twenty-first Annual Frederick William Reynolds Lecture, *Bulletin of the University of Utah* 48.

Nelson, Lowry L. 1952. *The Mormon Village: A Pattern and Technique of Land Settlement.* Salt Lake City: Univ. of Utah Press.

Olsen, Steven L. 1996–97. "Celebrating Cultural Identity: Pioneer Day in Nineteenth-Century Mormonism." *BYU Studies* 36, no. 1: 159–77.

Ostling, Richard N., and Joan K. Ostling. 1999. *Mormon America: The Power and the Promise.* San Francisco: Harper.

O'Dea, Thomas F. 1957. *The Mormons.* Chicago: Univ. of Chicago Press.

Robinson, Philip S. 1883. *Sinners and Saints.* Boston: Roberts Bros.

Sauder, Robert A. 1996. "State v. Society: Public Land Law and Mormon Settlement in the Sevier Valley, Utah." *Agricultural History* 70, no. 1: 57–89.

Shipps, Jan. 1985. *Mormonism: The Story of a New Religious Tradition.* Urbana: Univ. of Illinois Press.

Shipps, Jan. 1994. "Making Saints: In the Early Days and the Latter Days." In *Contemporary Mormonism: Social Science Perspectives,* ed. Marie Cornwall et al., 64–83. Urbana: Univ. of Illinois Press.

Shipps, Jan. 1998. "Differences and Otherness: Mormonism and the American Religious Mainstream." In *Minority Faiths and the American Protestant Mainstream,* ed. Jonathan D. Sarna, 81–109. Urbana: Univ. of Illinois Press.

Sillitoe, Linda. 1996. *Salt Lake County History*. Salt Lake City: Utah State Historical Society.

Smart, William B., and John Telford. 1995. *Utah: A Portrait*. Salt Lake City: Univ. of Utah Press.

Stegner, Wallace. 1969. *The Sound of Mountain Water*. Garden City, N.Y.: Doubleday.

Stegner, Wallace. 1942. *Mormon Country*. Reprint, 1981. Lincoln: Univ. of Nebraska Press.

Taniguchi, Nancy J. 1996. *Necessary Fraud: Progressive Reform and Utah Coal*. Norman: Univ. of Oklahoma Press.

Taylor, Samuel W. 1978. *Rocky Mountain Empire: The Latter-day Saints Today*. New York: Macmillan.

Thayne, Emma Lou. 1998. "Through the Viewer: This Is My Place." In *New Genesis: A Mormon Reader on Land and Community*, ed. Terry Tempest Williams, William B. Smart, and Gibbs Smith, 251–57. Layton, Utah: Gibbs Smith.

Thompson, Brent G. 1983. "'Standing between Two Fires': Mormons and Prohibition, 1908–1917." *Journal of Mormon History* 10: 35–52.

Webster, Donovan. 1996. "Utah: Land of Promise, Kingdom of Stone." *National Geographic*, January, 48–77.

Wright, John B. 1998. *Montana Ghost Dance: Essays on Land and Life*. Austin: Univ. of Texas Press.

CHAPTER 13: *California's Emerging Russian Homeland*

Alexseov, Mikhail. 1985. Interview with Hardwick. San Francisco, 23 December.

Bolyshkanov, Nadia. 1998. Personal correspondence with Hardwick. Rancho Cordova, 17 April.

California Department of Refugee Services. 1991. *Estimates of Refugees in California Counties and the State*. Sacramento: State of California.

Domasky, Pauline. 1983. Interview with Hardwick. West Sacramento, 2 March.

Essig, E.O. 1933. "The Russian Colonies in California." *Quarterly of the California Historical Society* 12, no. 2: 189–90.

Gans, Herbert J. 1962. *The Urban Villagers: Group and Class in the Life of Italian Americans*. New York: Free Press of Glencoe.

Gibson, James R. 1976. *Imperial Russia in Frontier America*. New York: Oxford Univ. Press.

Hardwick, Susan Wiley. 1986. "Ethnic Residential and Commercial Patterns in Sacramento with Special Reference to the Russian-American Experience." Ph.D. diss., Univ. of California, Davis.

Hardwick, Susan Wiley. 1993. *Russian Refuge: Religion, Migration, and Settlement on the North American Pacific Rim*. Chicago: Univ. of Chicago Press.

Kantor, James R. K., ed. 1880. *Grimshaw's Narrative*. Orig. published in Thompson and West's *History of Sacramento County*. Sacramento: Sacramento Book Collector's Club.

Karakozoff, David. 1985. Interview with Hardwick. Sacramento, 1 November.

Kondratieff, George. 1976. "Fiftieth Anniversary of the Holy Myrrhbearer's." *American Orthodox Church in Bryte, California.* Bryte, Calif.: Orthodox Church Press.

Larkey, Joanne L., and Shipley Walters. 1987. *Yolo County: Land of Changing Patterns.* Northridge, Calif.: Windsor Publications.

Lokteff, Mikhail. 1984. Interview with Hardwick. West Sacramento, 10 March.

Morgunov, Paul. 1990. Interview with Hardwick. Chico, 11 August.

Nordyke, Eleanor. 1977. *The Peopling of Hawaii.* Honolulu: East-West Center.

Planteen, Nina. 1985. Interview with Hardwick. West Sacramento, 19 September.

Tuan, Yi-Fu. 1974. *Topophilia: A Study of Environmental Perception, Attitudes, and Values.* Englewood Cliffs, N.J.: Prentice-Hall.

Tuan, Yi-Fu. 1977. *Space and Place: The Perspective of Experience.* Minneapolis: Univ. of Minnesota Press.

Vance, James E. 1976. "The American City: Workshop for a National Culture." In *Contemporary Metropolitan America.* Washington, D.C.: Association of American Geographers.

Vesselerova, Louisa. 1985. Interview with Hardwick. West Sacramento, 22 October.

CHAPTER 14: *Montana's Emerging Montane Homeland*

DeVoto, Bernard. 1934. "The West: A Plundered Province." *Harper's Monthly Magazine,* August, 355–64.

Francaviglia, Richard V. 1970. "The Mormon Landscape: Definition of an Image of the American West." *Proceedings of the Association of American Geographers* 1: 59–61.

Garreau, Joel. 1981. *The Nine Nations of North America.* New York: Avon Books.

Howard, Joseph Kinsey. 1943. *Montana: High, Wide, and Handsome.* Lincoln: Univ. of Nebraska Press.

Jordan, Terry G. 1993. *North American Cattle-Ranching Frontiers: Origins, Diffusion, and Differentiation.* Albuquerque: Univ. of New Mexico Press.

Kittredge, William, and Annick Smith. 1988. *The Last Best Place: A Montana Anthology.* Helena: Montana Historical Society.

Limerick, Patricia Nelson. 1988. *The Legacy of Conquest: The Unbroken Past of the American West.* New York: Norton.

Meinig, D. W. 1972. "American Wests: Preface to a Geographical Interpretation." *Annals of the Association of American Geographers.* 62: 159–84.

Montana Supreme Court. 1999. *Montana Environmental Information Center, et al. v. Department of Environmental Quality and 7–Up Pete Joint Venture.* MT 248. Helena, Mont.

Opie, John. 1993. *Ogallala: Water for a Dry Land.* Lincoln: Univ. of Nebraska Press.

Peirce, Neal R. 1972. *The Mountain States of America: People, Politics, and Power in the Eight Rocky Mountain States.* New York: Norton.

Starrs, Paul F. 1998. *Let the Cowboy Ride: Cattle Ranching in the American West.* Baltimore: Johns Hopkins Univ. Press.

Starrs, Paul F., and John B. Wright. 1995. "Great Basin Growth and the Withering of California's Pacific Idyll." *Geographical Review* 85: 417–35.

Toole, K. Ross. 1959. *Montana: An Uncommon Land.* Norman: Univ. of Oklahoma Press.

Toole, K. Ross. 1972. *Twentieth-Century Montana: A State of Extremes.* Norman: Univ. of Oklahoma Press.

Turner, Frederick Jackson. 1920. *The Frontier in American History.* New York: Henry Holt.

Von Reichert, Christiane. 1998a. "Montana's Changing Migration Patterns." Bureau of Business and Economic Research. Research Paper. Univ. of Montana, Missoula.

Von Reichert, Christiane. 1998b. "Who Are the Migrants and Why Do They Come?" Bureau of Business and Economic Research. Research Paper. Univ. of Montana, Missoula.

Von Reichert, Christiane, and James T. Sylvester. 1998. "Motives for Migration: A Study of Montana Newcomers." *Montana Business Quarterly*, Winter, 15–19.

Webb, Walter Prescott. 1957. "The American West, Perpetual Mirage." *Harper's Magazine* 214, no. 1284: 25–31.

Wright, John B. 1998. *Montana Ghost Dance: Essays on Land and Life.* Austin: Univ. of Texas Press.

Zelinsky, Wilbur. 1992. *The Cultural Geography of the United States.* Revised Edition. Englewood Cliffs, N.J.: Prentice-Hall.

CHAPTER 15: *American Homelands: A Dissenting View*

Adler, Joseph. 1997. *Restoring the Jews to Their Homeland: Nineteen Centuries in the Quest for Zion.* Northvale, N.J.: J. Aronson.

Ames, Merlin. M. 1939. *Homelands: America's Old-World Backgrounds.* St. Louis: Webster Publishing Co.

Anderson, James. 1988. "Nationalist Ideology and Territory." In *Nationalism, Self-Determination and Political Geography*, ed. Ron Johnston, David B. Knight, and Eleonore Kofman, 18–39. New York: Croom Helm.

Arreola, Daniel D. 1987. "The Mexican American Cultural Capital." *Geographical Review* 77, no. 1: 17–34.

Arreola, Daniel D. 1993. "The Texas-Mexican Homeland." *Journal of Cultural Geography* 13, no. 2: 61–74.

Barkan, Elliott R. 1995. "Race, Religion, and Nationality in American Society: A Model of Ethnicity—From Contact to Assimilation." *Journal of American Ethnic History* 14, no. 2: 38–75.

Bennion, Lowell C. 1980. "Mormon Country a Century Ago: A Geographer's View."

In *The Mormon People*, ed. Thomas G. Alexander, 1–26. Provo: Brigham Young Univ. Press.

Bjorklund, Elaine M. 1992. "The Namerind Continent's Euro-American Transformation, 1600–1950." In *Geographical Snapshots of North America*, ed. Donald G. Janelle, 5–10. New York: Guilford Press.

Boswell, Thomas D. 1993. "The Cuban-American Homeland in Miami." *Journal of Cultural Geography* 13, no. 2: 133–48.

Campbell, John C. 1921. *The Southern Highlander and His Homeland*. New York: Russell Sage Foundation.

Carlson, Alvar W. 1990. *The Spanish-American Homeland: Four Centuries in New Mexico's Río Arriba*. Baltimore: Johns Hopkins Univ. Press.

Cohen, Robin. 1997. *Global Diasporas: An Introduction*. Seattle: Univ. of Washington Press.

Comeaux, Malcolm L. 1977. "Les Acadiens Louisianais: L'impact de l'environnement." *Revue de Louisiane/Louisiana Review* 6, no. 2: 163–78.

Comeaux, Malcolm L. 1996a. "Cajuns and Their Adaptation to a Modern World." In *Human Geography in North America: New Perspectives and Trends in Research*, ed. Klaus Frantz, 7–16. Innsbruck: Universität Innsbruck, Innsbrucker Geographische Studien, vol. 26.

Comeaux, Malcolm L. 1996b. "The Acadians of Canada and the Cajuns of Louisiana: Cultural Change over Time and Distance." In *Ethnic Persistence and Change in Europe and America: Traces in Landscape and Society*, ed. Klaus Frantz and Robert A. Sauder, 29–45. Innsbruck: Veröffentlichungen der Universität Innsbruck, vol. 213.

Connor, Walker. 1978. "A Nation is a Nation, is a State, is an Ethnic Group is a . . ." *Ethnic and Racial Studies* 1: 77–400.

Connor, Walker. 1986. "The Impact of Homelands upon Diasporas." In *Modern Diasporas in International Politics*, ed. Gabriel Sheffer, 16–46. New York: St. Martin's Press.

Conzen, Michael P. 1993. "Culture Regions, Homelands, and Ethnic Archipelagos: Methodological Considerations." *Journal of Cultural Geography* 13, no. 2: 13–29.

Conzen, Michael P. 1996. "The German-Speaking Ethnic Archipelago in America." In *Ethnic Persistence and Change in Europe and America: Traces in Landscape and Society*, ed. Klaus Frantz and Robert A. Sauder, 71–96. Innsbruck: Veröffentlichungen der Universität Innsbruck, vol. 213.

Conzen, Michael P., ed. 1990. *The Making of the American Landscape*. Boston: Unwin Hyman.

Crowley, William K. 1978. "Old Order Amish Settlement: Diffusion and Growth." *Annals of the Association of American Geographers* 68, no. 2: 249–64.

Davis, George A., and O. Fred Donaldson. 1975. *Blacks in the United States: A Geographic Perspective*. Boston: Houghton Mifflin.

Dockendorff, Thomas P. 1986. "Cultural Geography of Eastern Stearns County: God, Granite, and Germans." In *AAG '86 Twin Cities Field Trip Guide*, ed. Thomas J.

Baerwald and Keith L. Harrington, 184–93. Washington, D.C.: Association of American Geographers.

Drachler, Jacob, comp. 1975. *Black Homeland/Black Diaspora: Cross-Currents of the African Relationship.* Port Washington, N.Y.: Kennikat Press.

Ehringhaus, Sibylle. 1996. *Germanenmythos und Deutsche Identität: Die Frühmittelalter-Rezeption in Deutschland 1842–1933.* Weimar: Verlag und Datenbank für Geisteswissenschaften.

Estaville, Lawrence E., Jr. 1986. "Mapping the Louisiana French." *Southeastern Geographer* 26, no. 2: 90–113.

Estaville, Lawrence E., Jr. 1988. "Were the Nineteenth-Century Cajuns Geographically Isolated?" in *The American South*, ed. Richard L. Nostrand and Sam B. Hilliard, 85–96. Geoscience and Man, vol. 25. Baton Rouge: Louisiana State Univ. School of Geoscience.

Estaville, Lawrence E., Jr. 1993. "The Louisiana-French Homeland." *Journal of Cultural Geography* 13, no. 2: 31–45.

Francaviglia, Richard V. 1978. *The Mormon Landscape: Existence, Creation, and Perception of a Unique Image in the American West.* New York: AMS Press.

Frantz, Klaus. 1999. *The Indian Reservations of the United States.* Chicago: Univ. of Chicago Press.

Gleason, Philip. 1980. "American Identity and Americanization." In *Harvard Encyclopedia of American Ethnic Groups*, ed. Stephan Thernstrom, 31–58. Cambridge, Mass.: Harvard Univ. Press.

Golab, Zbigniew. 1992. *The Origins of the Slavs: A Linguist's View.* Columbus, Ohio: Slavica Publications.

Helzer, Jennifer Jill. 1998. "The Italian Ethnic Substrate in Northern California: Cultural Transfer and Regional Identity." Ph.D. diss., University of Texas at Austin.

Herzl, Theodor, et al. 1989. "Historical Debate: Is a Homeland for the Jews Necessary?" In *Israel: Opposing Viewpoints*, ed. Janelle Rohr and Bob Anderson, 16–49. San Diego, Calif.: Greenhaven Press.

Hobsbawm, Eric. 1990. *Nations and Nationalism since 1780: Programme, Myth, Reality.* New York: Cambridge Univ. Press.

Jordan, Terry G. 1993a. "The Anglo-Texan Homeland." *Journal of Cultural Geography* 13, no. 2: 75–86.

Jordan, Terry G. 1993b. "The Anglo-American Mestizos and Traditional Southern Regionalism." In *Culture, Form, and Place: Essays in Cultural and Historical Geography*, ed. Kent Mathewson, 175–95. Geoscience and Man, vol. 32. Baton Rouge: Louisiana State Univ. School of Geoscience.

Jordan, Terry G., and Matti E. Kaups. 1989. *The American Backwoods Frontier: An Ethnic and Ecological Interpretation.* Baltimore: Johns Hopkins Univ. Press.

Kaiser, Robert J. 1994. "The Meaning of Homeland in the Study of Nationalism." *The Geography of Nationalism in Russia and the USSR*, 3–32. Princeton, N.J.: Princeton Univ. Press.

Kivisto, Peter. 1987. "Finnish Americans and the Homeland, 1918–1958." *Journal of American Ethnic History* 7, no. 1: 7–28.

Knight, David B. 1982. "Identity and Territory: Geographical Perspectives on Nationalism and Regionalism." *Annals of the Association of American Geographers* 72, no. 4: 514–31.

Lamme, Ary J., III, and Douglas B. McDonald. 1993. "The 'North Country' Amish Homeland." *Journal of Cultural Geography* 13, no. 2: 107–18.

Limouze, Arthur H. 1939. *Homeland Harvest.* New York: Friendship Press.

Lipton, Merle, and Charles E. W. Simkins, eds. 1993. *State and Market in Post-Apartheid South Africa.* Boulder: Westview Press.

Manczak, W. 1986. "The Original Homeland of the Goths." *Folia Linguistica Historica* 7, no. 2: 371–80.

Markale, Jean. 1993. *The Celts: Uncovering the Mythic and Historic Origins of Western Culture.* Rochester, Vt.: Inner Traditions.

May, Dean L. 1994. "The State of Deseret." In *Historical Atlas of Mormonism,* ed. S. Kent Brown et al., 90–91. New York: Simon & Schuster.

Meinig, Donald W. 1965. "The Mormon Culture Region: Strategies and Patterns in the Geography of the American West, 1847–1967." *Annals of the Association of American Geographers* 55, no. 2: 191–220.

Meinig, Donald W. 1969. *Imperial Texas: An Interpretive Essay in Cultural Geography.* Austin: Univ. of Texas Press.

Meinig, Donald W. 1971. *Southwest: Three Peoples in Geographical Change, 1600–1970.* New York: Oxford Univ. Press.

Mikesell, Marvin W. 1983. "The Myth of the Nation State." *Journal of Geography* 82, no. 6: 257–60.

Mitchell, Robert D. 1978. "The Formation of Early American Cultural Regions: An Interpretation." In *European Settlement and Development in North America: Essays in Geographical Change in Honour and Memory of Andrew Hill Clark,* ed. James R. Gibson, 66–90. Toronto: Univ. of Toronto Press.

Mitchell, Robert D., ed. 1991. *Appalachian Frontiers: Settlement, Society, and Development in the Pre-Industrial Era.* Lexington: Univ. Press of Kentucky.

Neumeyer, Michael. 1992. *Heimat: Zu Geschichte und Begriff eines Phänomens.* Kiel: Geographische Institut der Universität Kiel, Kieler Geographische Schriften, vol. 84.

Nobutaka, Inoue. 1997. *Globalization and Indigenous Culture.* Tokyo: Kokugakuin Univ., Institute for Japanese Culture and Classics.

Nostrand, Richard L. 1992. *The Hispano Homeland.* Norman: Univ. of Oklahoma Press.

Nostrand, Richard L., and Lawrence E. Estaville, Jr. 1993. "Introduction: The Homeland Concept." *Journal of Cultural Geography* 13, no. 2: 1–4.

Nostrand, Richard L., and Lawrence E. Estaville, Jr., eds. 1995. Transcript of Panel Discussion: The Homeland Concept, American Ethnic Group Specialty Group, Association of American Geographers 1994 Annual Meeting (San Francisco).

O'Leary, Timothy J., and David Levinson, eds. 1991. *Encyclopedia of World Cultures*, vol. 1: *North America*. Boston: G. K. Hall.

Petersen, William. 1980. "Concepts of Ethnicity." In *Harvard Encyclopedia of American Ethnic Groups*, ed. Stephan Thernstrom, 31–58. Cambridge, Mass.: Harvard Univ. Press.

Poche, Bernard. 1992. "Identification as a Process: Territories as an Organizational or a Symbolic Area." In *Globalization and Territorial Identities*, ed. Zdravko Mlinar, 129–49. Brookfield, Vt.: Avebury.

Prucha, Francis Paul. 1990. *Atlas of American Indian Affairs*, fig. 3: Indian Land Areas Judicially Established, 6–7. Lincoln: Univ. of Nebraska Press.

Raitz, Karl B. 1978. "North American Ethnic Maps." *Geographical Review* 68, no. 3: 335–50.

Roark, Michael O. 1993. "Homelands: A Conceptual Essay." *Journal of Cultural Geography* 13, no. 2: 5–11.

Roosens, Eugeen E. 1989. *Creating Ethnicity: The Process of Ethnogenesis*. Frontiers of Anthropology, vol. 5. Newberry Park, Calif.: Sage Publications.

Ross, Thomas E., and Tyrel G. Moore, eds. 1987. *A Cultural Geography of North American Indians*. Boulder: Westview Press.

Sack, Kevin. 1997. "Symbols of Old South Feed a New Bitterness." *New York Times*, 8 February, 1 and 8.

Sack, Robert. 1986. *Human Territoriality: Its Theory and History*. Cambridge: Cambridge Univ. Press.

Schlesinger, Arthur, Jr. 1992. *The Disuniting of America*. New York: Norton.

Shell, Marc. 1993. "Babel in America; or, the Politics of Language Diversity in the United States." *Critical Inquiry* 20, no. 1: 103–27.

Sheskin, Ira M. 1993. "Jewish Metropolitan Homelands." *Journal of Cultural Geography* 13, no. 2: 119–32.

Siegel, Jacob. 1996. "Geographic Compactness vs. Race/Ethnic Compactness and Other Criteria in the Delineation of Legislative Districts." *Population Research and Policy Review* 15, no. 2: 147–64.

Smith, Anthony D. 1992. "Chosen Peoples: Why Ethnic Groups Survive." *Ethnic and Racial Studies* 15, no. 3: 436–56.

Sutton, Imre, ed. 1985. *Irredeemable America: The Indians' Estate and Land Claims*. Albuquerque: Univ. of New Mexico Press.

Tesón, Fernando R. 1998. "Ethnicity, Human Rights, and Self-Determination." In *International Law and Ethnic Conflict*, ed. David Wippman, 86–111. Ithaca, N.Y.: Cornell Univ. Press.

Thernstrom, Stephan, ed. 1980. *Harvard Encyclopedia of American Ethnic Groups*. Cambridge, Mass.: Harvard Univ. Press.

Thurston, Ernest L., and Grace C. Hankins. 1954. *Homelands of the Americas*. Syracuse, N.Y.: Iroquois Publishing Co.

Tuan, Yi-Fu. 1977. "Attachment to Homeland." *Space and Place: The Perspective of Experience*, 149–60. Minneapolis: Univ. of Minnesota Press.

Vassady, Béla. 1982. "The Homeland Cause as Stimulant to Ethnic Identity: The Hungarian-American Response to Károlyis 1914 American Tour." *Journal of American Ethnic History* 2, no. 1: 39–64.

Vogeler, Ingolf K. 1976. "The Roman Catholic Culture Region of Central Minnesota." *Pioneer America* 8, no. 2: 71–83.

Waddell, Eric. 1983. "French Louisiana: An Outpost of l'Amérique Française or Another Country and Another Culture?" in *French America: Mobility, Identity, and Minority Experience across the Continent*, ed. Dean R. Louder and Eric Waddell, 229–51. Baton Rouge: Louisiana State Univ. Press.

Zelinsky, Wilbur. 1973. *The Cultural Geography of the United States*. Englewood Cliffs, N.J.: Prentice-Hall.

Zelinsky, Wilbur. 1988. *Nation into State: The Shifting Symbolic Foundations of American Nationalism*. Chapel Hill: Univ. of North Carolina Press.

Zelinsky, Wilbur. 1990. "Seeing beyond the Dominant Culture." *Places: A Quarterly Journal of Environmental Design* 7, no. 1: 32–35.

Zelinsky, Wilbur. 1992. "The American Culture Area Revisited." *The Cultural Geography of the United States*, rev. ed., 177–82. Englewood Cliffs, N.J.: Prentice-Hall.

Zelinsky, Wilbur, and Barrett A. Lee. 1998. "Heterolocalism: An Alternative Model of the Sociospatial Behavior of Immigrant Ethnic Communities." *International Journal of Population Geography* 4, no. 1: 1–18.

CONTRIBUTORS

CHARLES S. AIKEN is professor of geography and a member of the American Studies faculty at the University of Tennessee, Knoxville. He is past president of the Southeastern Division, Association of American Geographers. Aiken's specialities are rural geography and geography of the American South. He is the author of numerous articles about the South and of *The Cotton Plantation South since the Civil War* (Johns Hopkins University Press, 1998), which received the 1999 J. B. Jackson Prize of the Association of American Geographers.

DANIEL D. ARREOLA is professor of geography and an affiliated faculty in the Center for Latin American Studies at Arizona State University. He specializes in the study of cultural landscapes in the Mexican American borderlands. His book co-authored with James R. Curtis, *The Mexican Border Cities: Landscape Anatomy and Place Personality* (University of Arizona Press, 1993), won a Southwest Book Award. In 1997–98 he served as president of the Association of Pacific Coast Geographers.

LOWELL C. "BEN" BENNION is professor emeritus of geography at Humboldt State University. Thanks to Donald Meinig and Leonard Arrington, he became involved in Mormon studies. He is the author of numerous articles on the geographic dynamics of Mormondom and co-author (with Gary B. Peterson) of *Sanpete Scenes: A Guide to Utah's Heart* (Basin/Plateau Press, 1987), which won the J. B. Jackson Prize of the Association of American Geographers in 1989, and (with Jerry Rohde) of *Traveling the Trinity Highway* (MountainHome Books, 2000).

MARTYN J. BOWDEN is professor of geography in the Graduate School of Geography, Clark University. He specializes in cultural, historical, humanistic, and urban geography and has many publications on cognition of the Great Plains and the American West, inner-city development (notably in San Francisco), and geographical change in New England. He founded the Eastern Historical Geography Association in 1969 and he was the founding editor of *Historical Geography* in 1971.

MICHAEL P. CONZEN is professor of geography at the University of Chicago. His interests include American historical and urban geography, landscape history, and the history of American commercial mapping. He has written extensively on America's evolving urban system, urban morphology, illustrated county atlases, and the Illinois & Michigan Canal region. His publications include *The Making of the American Landscape* (Unwin Hyman, 1990) and *A Scholar's Guide to Geographical Writing on the American and Canadian Past* (University of Chicago Press, 1993). The Austrian Academy of Sciences in Vienna recently elected him a corresponding member.

LAWRENCE E. ESTAVILLE is professor of geography and chairman of the Department of Geography at Southwest Texas State University. His primary research focus is the geographical experience of the Louisiana French, especially the Cajuns. His *Confederate Neckties: Louisiana Railroads in the Civil War* (McGinty Publications, Louisiana Tech University, 1989) won the Historic New Orleans Collection's L. Kemper Williams Prize.

SUSAN W. HARDWICK is associate professor of geography at the University of Oregon. Her research interests concern ethnic migration and settlement in western North America and geographic education. The author of numerous journal articles and scholarly books on the Russian immigrant experience, Susan is best known for her *Russian Refuge: Religion, Migration, and Settlement on the Pacific Rim of North America* (University of Chicago Press, 1993) and *Geography for Educators: Standards, Themes and Concepts* (Prentice-Hall, 1996).

STEPHEN C. JETT is professor emeritus of geography and former chairman of the late Department of Geography, University of California, Davis. His research specialties are American Indians of the American Southwest, especially the Navajo, and possible pre-Columbian transoceanic contacts between the hemispheres. He is the author of the award-winning books *Navajo Wildlands* (Sierra Club, 1967, with photographs by Philip Hyde) and (with Virginia E. Spencer) *Navajo Architecture: Forms, History, Distributions* (University of Arizona Press, 1981). He edits the new journal *Pre-Columbiana: A Journal of Long-Distance Contacts.*

TERRY G. JORDAN-BYCHKOV holds the Walter Prescott Webb Chair in Geography at the University of Texas at Austin. Major scholarly publications include *The Mountain West: Interpreting the Folk Landscape* (Johns Hopkins University Press, 1997), *North American Cattle Ranching Frontiers* (University of New Mexico Press, 1993), *The American Backwoods Frontier* (Johns Hopkins University Press, 1989), and *American Log Buildings* (University of North Carolina Press, 1984). He served as president of the Association of American Geographers in 1987–88. Honors include membership in the Texas Institute of Letters, election as a Fellow of the Texas State Historical Association, and an Honors Award from the Association of American Geographers.

ARY J. LAMME III is associate professor of geography at the University of Florida. Graduate training at Syracuse University fostered his attempts to combine seemingly disparate elements in the search for holistic geographic explanation. These elements include patterns of religious adherence, landscape perception and utilization, value orientation in space, and varying representational methods. His *America's Historic Landscapes* (University of Tennessee Press, 1990) addresses these concerns in regard to historic preservation.

RICHARD L. NOSTRAND is David Ross Boyd Professor of Geography at the University of Oklahoma. His long-term interest in the changing geography of Spanish and Mexican Americans in the U.S. borderlands has focused recently on the concept of the homeland as a basis for cultural regionalism. His *Hispano Homeland* (University of Oklahoma Press, 1992) won the first Angie Debo Prize.

RICHARD PILLSBURY is interested in the evolution and character of the American landscape. His recent publications include *No Foreign Food: The American Diet in Time and Place* (Westview Press, 1998), *An Atlas of American Agriculture: The American Cornucopia* (Simon & Schuster Macmillan, 1996) with John Florin, and *From Boarding House to Bistro: The American Restaurant Then and Now* (Unwin & Hyman, 1990). He is professor emeritus of geography at Georgia State University.

STEVEN M. SCHNELL is assistant professor of geography at Northwest Missouri State University. His interest in Kiowas stems from having grown up in Norman, Oklahoma. This chapter is drawn from his Master's thesis, "The Kiowa Homeland in Oklahoma," written under James R. Shortridge at the University of Kansas. Schnell's dissertation at the University of Kansas concerns tourism, ethnicity, and identity in Lindsborg, Kansas.

JOHN B. WRIGHT is associate professor of geography at New Mexico State University. He specializes in the study and conservation of cultural landscapes in the American West. His book *Rocky Mountain Divide: Selling and Saving the West* (University of Texas Press, 1993) won the J. B. Jackson Prize and the Publication Award of the Geographic Society of Chicago. Other books include *Montana Ghost Dance: Essays on Land and Life* (University of Texas Press, 1998) and *Montana Places: Exploring Big Sky Country* (University of Minnesota Press, 2000). Wright teaches in New Mexico but owns a home in Missoula, Montana, where he spends every summer working as a conservation consultant to Montana land trust organizations.

INDEX

Acadia, 83, 84–85. *See also* Cajuns; Nouvelle Acadie
Africa, as homeland, 57–58
African Americans, xv, xix, 58, 88, 239, 247, 266. *See also* Blacks
agricultural practices, 85–86, 92, 108
Alabama, 63, 64
Alamo, xx, 127, 133
Albuquerque, N.Mex., 165, 206
Allegheny Front (Pa.), 39
American Indians. *See* Indians
Americanism, 261
Americanization, 21, 43, 100. *See also* mainstream America; national culture
Amish, xix, xxi, 42, 44–52, 253, 265, 268; bonding with place among, 46, 49, 52; community among, 49–51; control of place by, 48–49; cultural impress of, 46–48; land ownership by, 51. *See also* Pennsylvania Dutch
Amish homeland, 51–52
Ammann, Jacob, 45
Anaconda, Mont., 227, 236
Anadarko, Okla., 144, 147
Aneth, Utah, 182
Anglo-Americans: in Louisiana, 83, 88, 89, 91, 92, 94, 95, 97, 98; in New England, 1–23; in South Texas, 107, 157
Anglo-Texans, xx, 125–38; bonding with

place, 125; control of place, 125, 135
Arcata, Calif., 214
Arizona, 168, 179
Arkansas River, 139
Atchafalaya Swamp (La.), 85, 89
Atlanta, Ga., 61
attachment to place. *See* bonding with place
Austin, Stephen F., 133, 162

Bakersfield, Calif., 211, 216, 219
Balcones Escarpment (Tex.), 107, 108
Barboncito, 177, 178
Barreiro, Antonio, 158, 165
Bayou Lafourche, La., 89, 91, 97
Bayou Teche, La., 85, 89, 90, 93, 97
Big Bow, 149
Billings, Mont., 296
Birmingham, Ala., 61
Blackfoot River (Mont.), 233
Blackfoot Valley (Mont.), 236
Black Hills (S.Dak.), 139
Blackland string prairies (Tex.), 109
Blacks, 53–72, 227; bonding with place among, 56–57, 60–61, 67, 69, 72; control of place by, 68, 71. *See also* African Americans
Blanca Peak (Colo.), 173
Boise, Idaho 206

RELATED TITLES IN THE SERIES

*The Cotton Plantation South since
the Civil War*
Charles S. Aiken

*Petrolia: The Landscape of America's
First Oil Boom*
Brian C. Black

*The Spanish-American Homeland:
Four Centuries in New Mexico's
Río Arriba*
Alvar W. Carlson

*The Pennsylvania Barn: Its Origin,
Evolution, and Distribution in
North America*
Robert F. Ensminger

*Measure of Emptiness: Grain
Elevators in the American Landscape*
Frank Gohlke, with a concluding
essay by John C. Hudson

*The American Backwoods Frontier:
An Ethnic and Ecological
Interpretation*
Terry G. Jordan and Matti Kaups

*The Mountain West: Interpreting the
Folk Landscape*
Terry G. Jordan, Jon T. Kilpinen,
and Charles F. Gritzner

*Everyday Architecture of the
Mid-Atlantic: Looking at Buildings
and Landscapes*
Gabrielle M. Lanier and
Bernard L. Herman

*To Build in a New Land: Ethnic
Landscapes in North America*
Edited by Allen G. Noble

Belonging to the West
Eric Paddock

*Delta Sugar: Louisiana's Vanishing
Plantation Landscape*
John B. Rehder

*The Four-Cornered Falcon:
Essays on the Interior West and
the Natural Scene*
Reg Saner

*Let the Cowboy Ride: Cattle
Ranching in the American West*
Paul F. Starrs

*Alligators: Prehistoric Presence
in the American Landscape*
Martha A. Strawn

The New England Village
Joseph S. Wood